Advances in Industrial Control

For other titles published in this series, go to
www.springer.com/series/1412

Other titles published in this series:

Digital Controller Implementation and Fragility
Robert S.H. Istepanian and James F. Whidborne (Eds.)

Optimisation of Industrial Processes at Supervisory Level
Doris Sáez, Aldo Cipriano and Andrzej W. Ordys

Robust Control of Diesel Ship Propulsion
Nikolaos Xiros

Hydraulic Servo-systems
Mohieddine Mali and Andreas Kroll

Model-based Fault Diagnosis in Dynamic Systems Using Identification Techniques
Silvio Simani, Cesare Fantuzzi and Ron J. Patton

Strategies for Feedback Linearisation
Freddy Garces, Victor M. Becerra, Chandrasekhar Kambhampati and Kevin Warwick

Robust Autonomous Guidance
Alberto Isidori, Lorenzo Marconi and Andrea Serrani

Dynamic Modelling of Gas Turbines
Gennady G. Kulikov and Haydn A. Thompson (Eds.)

Control of Fuel Cell Power Systems
Jay T. Pukrushpan, Anna G. Stefanopoulou and Huei Peng

Fuzzy Logic, Identification and Predictive Control
Jairo Espinosa, Joos Vandewalle and Vincent Wertz

Optimal Real-time Control of Sewer Networks
Magdalene Marinaki and Markos Papageorgiou

Process Modelling for Control
Benoît Codrons

Computational Intelligence in Time Series Forecasting
Ajoy K. Palit and Dobrivoje Popovic

Modelling and Control of Mini-Flying Machines
Pedro Castillo, Rogelio Lozano and Alejandro Dzul

Ship Motion Control
Tristan Perez

Hard Disk Drive Servo Systems (2nd Ed.)
Ben M. Chen, Tong H. Lee, Kemao Peng and Venkatakrishnan Venkataramanan

Measurement, Control, and Communication Using IEEE 1588
John C. Eidson

Piezoelectric Transducers for Vibration Control and Damping
S.O. Reza Moheimani and Andrew J. Fleming

Manufacturing Systems Control Design
Stjepan Bogdan, Frank L. Lewis, Zdenko Kovačić and José Mireles Jr.

Windup in Control
Peter Hippe

Nonlinear H_2/H_∞ Constrained Feedback Control
Murad Abu-Khalaf, Jie Huang and Frank L. Lewis

Practical Grey-box Process Identification
Torsten Bohlin

Control of Traffic Systems in Buildings
Sandor Markon, Hajime Kita, Hiroshi Kise and Thomas Bartz-Beielstein

Wind Turbine Control Systems
Fernando D. Bianchi, Hernán De Battista and Ricardo J. Mantz

Advanced Fuzzy Logic Technologies in Industrial Applications
Ying Bai, Hanqi Zhuang and Dali Wang (Eds.)

Practical PID Control
Antonio Visioli

(*continued after Index*)

Fabrizio Caccavale · Mario Iamarino ·
Francesco Pierri · Vincenzo Tufano

Control and Monitoring of Chemical Batch Reactors

Springer

Prof. Fabrizio Caccavale
Dipartimento di Ingegneria e Fisica dell'Ambiente
Università degli Studi della Basilicata
Viale dell'Ateneo Lucano 10
85100 Potenza
Italy
fabrizio.caccavale@unibas.it

Francesco Pierri
Dipartimento di Ingegneria e Fisica dell'Ambiente
Università degli Studi della Basilicata
Viale dell'Ateneo Lucano 10
85100 Potenza
Italy
francesco.pierri@unibas.it

Mario Iamarino
Dipartimento di Ingegneria e Fisica dell'Ambiente
Università degli Studi della Basilicata
Viale dell'Ateneo Lucano 10
85100 Potenza
Italy
mario.iamarino@unibas.it

Vincenzo Tufano
Dipartimento di Ingegneria e Fisica dell'Ambiente
Università degli Studi della Basilicata
Viale dell'Ateneo Lucano 10
85100 Potenza
Italy
vincenzo.tufano@unibas.it

ISSN 1430-9491
ISBN 978-0-85729-194-3 e-ISBN 978-0-85729-195-0
DOI 10.1007/978-0-85729-195-0
Springer London Dordrecht Heidelberg New York

British Library Cataloguing in Publication Data
A catalogue record for this book is available from the British Library

Cover design: eStudio Calamar S.L.

Printed on acid-free paper

Springer is part of Springer Science+Business Media (www.springer.com)

Advances in Industrial Control

Professor (Emeritus) O.P. Malik
Department of Electrical and Computer Engineering
University of Calgary
2500, University Drive, NW
Calgary, Alberta
T2N 1N4
Canada

Professor K.-F. Man
Electronic Engineering Department
City University of Hong Kong
Tat Chee Avenue
Kowloon
Hong Kong

Professor G. Olsson
Department of Industrial Electrical Engineering and Automation
Lund Institute of Technology
Box 118
221 00 Lund
Sweden

Professor A. Ray
Department of Mechanical Engineering
Pennsylvania State University
0329 Reber Building
University Park
PA 16802
USA

Professor D.E. Seborg
Chemical Engineering
University of California Santa Barbara
3335 Engineering II
Santa Barbara
CA 93106
USA

Doctor K.K. Tan
Department of Electrical and Computer Engineering
National University of Singapore
4 Engineering Drive 3
Singapore 117576
Singapore

Professor I. Yamamoto
Department of Mechanical Systems and Environmental Engineering
Faculty of Environmental Engineering
The University of Kitakyushu
1-1, Hibikino, Wakamatsu-ku, Kitakyushu, Fukuoka, 808-0135
Japan

To our families.

Series Editors' Foreword

The series *Advances in Industrial Control* aims to report and encourage technology transfer in control engineering. The rapid development of control technology has an impact on all areas of the control discipline. New theory, new controllers, actuators, sensors, new industrial processes, computer methods, new applications, new philosophies..., new challenges. Much of this development work resides in industrial reports, feasibility study papers and the reports of advanced collaborative projects. The series offers an opportunity for researchers to present an extended exposition of such new work in all aspects of industrial control for wider and rapid dissemination.

The broader objectives of process control engineering include:

(i) controlling processes and technology safely, thereby protecting process operators and workers and the natural environment
(ii) minimizing the energy resources required to operate the process (in a wider environmental context, this also reduces the need to generate and deliver more energy to the process); and
(iii) operating the process or technology to optimize the material resource consumption (one aspect of this optimization is the simple reduction in the quantity of material used, but another is to use the same quantity of material to produce more consistent and better quality end products).

An interesting feature of these objectives is that they transcend application domains, applying as well to the new emerging technologies being devised to ensure future sustainability as to the traditional technological processes of industrial control. Thus, the real strength of industrial control engineering science lies in the universality of its techniques across application and industrial domains.

This *Advances in Industrial Control* monograph, *Control and Monitoring of Chemical Batch Reactors*, by Fabrizio Caccavale, Mario Iamarino, Francesco Pierri and Vincenzo Tufano exemplifies this universality extremely well. The domain of application, the chemical batch reactor, is part of chemical and process engineering; the process objectives are safe process operation, minimal energy consumption, and enhanced quality and consistency of operation. The roadmap of this study of a mature technology is in four stages:

(i) process modelling
(ii) model parameter identification
(iii) control design, simulation and verification; and
(iv) analysis for a fault-handling system.

The monograph reports the stages in a very systematic manner and uses the phenol–formaldehyde reaction as a thematic case study throughout. Thus, chemical, process and control engineers can follow the general control framework and then see the authors' ideas in action using the case study process. In reporting the control design (Chap. 5), the widely used industrial structure of a cascade two-loop structure is employed, but the controllers exploit the model information from earlier chapters to give a nonlinear control scheme that incorporates adaptation. Next, the monograph reports the development of a fault detection and isolation (FDI) system (Chap. 6). The inclusion of the considerations for a FDI system is rarer in this kind of study, but here it is a demonstration of the value of the full four-part control system development roadmap.

This monograph will appeal to a wide readership. Industrial chemical and process engineers wishing to understand the application of modern control system ideas and the potential of nonlinear control more comprehensively will find much to study. The research community of control academics and postgraduate students will appreciate the interaction between the science of control engineering and the demanding control problems of batch reactors. They should find the application of the techniques to the case study a source of inspiration for future research. The monograph is a valuable addition to the *Advances in Industrial Control* series.

Readers from the fields of process, chemical and control engineering may find these monographs from the *Advances in Industrial Control* series of complementary interest: *Fault-tolerant Control Systems* by Hassan Noura, Didier Theilliol, Jean-Christophe Ponsart and Abbas Chamseddine (ISBN 978-1-84882-652-6, 2009); *Predictive Functional Control* by Jacques Richalet and Donal O'Donovan (ISBN 978-1-84882-492-8, 2009); and *Process Control* by Jie Bao and Peter L. Lee (ISBN 978-1-84628-892-0, 2007).

From the Editors' sister series, *Advanced Textbooks in Control and Signal Processing*, the volume *Analysis and Control of Nonlinear Process Systems* by Katalin M. Hangos, Jósef Bokor and Gábor Szederkényi (ISBN 978-1-85233-600-4, 2003) is also focussed on process control and the design of nonlinear controllers.

Industrial Control Centre
Glasgow
Scotland, UK

M.J. Grimble
M.A. Johnson

Preface

Batch chemical processes are widely used in the production of fine chemicals, pharmaceutical products, polymers, and many other materials. Moreover, the flexibility of batch processes has become an attractive feature because of the actual turbulence of markets, characterized by a rapidly changing demand.

Batch processes are often nonisothermal and characterized by nonlinear dynamics, whose effects are further emphasized by intrinsically unsteady operating conditions. Hence, methodological and technological problems related to batch chemical reactors are often very challenging and require contributions from different disciplines (chemistry, chemical engineering, control engineering, measurement, and sensing).

A number of issues need to be resolved when dealing with batch reactors in industrial applications, ranging from design and planning of the plant to scheduling, optimization, and performance achievement of batch operations. Performance is usually specified in terms of productivity of the plant, safety of operations, and quality of final products. In order to meet such requirements, several problems need to be addressed:

- modeling the reactor and the process
- identification of the parameters in the mathematical models
- control of the state variables characterizing the process; and
- early diagnosis of failures and faults accommodation.

This book is aimed at tackling the above problems from a joint academic and industrial perspective. Namely, advanced solutions (i.e., based on recent research results) to the four fundamental problems of modeling, identification, control, and fault diagnosis are developed in detail in seven chapters.

In each chapter, a general overview of foundational concepts is given, together with a review of classical and recent literature related to the various topics covered. In detail, the first chapter provides a comprehensive introduction to the main topics of the book, whereas the last chapter presents some suggestions for future research activity in this field.

The second chapter presents an introduction to modeling techniques of batch chemical reactors, with a particular emphasis on chemical kinetics. The third chapter provides a general introduction to the problem of identification of mathematical models; the general methodologies are reviewed and developed in a form suitable for identifying kinetic models of chemical reactions taking place in batch reactors. In the fourth chapter, the mathematical modeling is extended to consider the thermal stability of batch reactors, thus providing a bridge towards the problems discussed in the following two chapters.

In the fifth chapter, a general overview of temperature control for batch reactors is presented; the focus is on model-based control approaches, with a special emphasis on adaptive control techniques. Finally, the sixth chapter provides the reader with an overview of the fundamental problems of fault diagnosis for dynamical systems, with a special emphasis on model-based techniques (i.e., based on the so-called *analytical redundancy* approach) for nonlinear systems; then, a model-based approach to fault diagnosis for chemical batch reactors is derived in detail, where both sensors and actuators failures are taken into account.

In order to provide a unitary treatment of the different topics and to give a firm link to the underlying practical applications, a common case study is developed through the course of the book. Namely, a batch process of industrial interest, i.e., the phenol-formaldehyde reaction for the production of phenolic resins, is adopted to test the modeling, identification, control, and diagnosis approaches developed in the book. In this way, a *roadmap* for the development of control and diagnosis systems is provided, ranging from the early phases of the process setting to the design of an effective control and diagnosis system.

In conclusion, the aim of the book is twofold:

- to bring to the attention of process engineers industrially feasible model-based solutions to control and diagnosis problems for chemical batch reactors, where such solutions in industrial contexts are often considered not feasible; and
- to disseminate recent results on nonlinear model-based control and diagnosis among researchers in the field of chemical engineering and process control, so as to stimulate further advances in the industrial applications of such approaches.

Hence, the book is directed to both industrial practitioners and academic researchers, although it is also suitable for adoption in advanced post-graduate level courses focused on process control, control applications, and nonlinear control.

Potenza

Fabrizio Caccavale, Mario Iamarino
Francesco Pierri, Vincenzo Tufano

Acknowledgements

The authors wish to thank Prof. M. Mattei and Dr. G. Paviglianiti, who collaborated to the development of some fault diagnosis schemes presented in the sixth chapter. Moreover, the authors are grateful to their former student G. Satriano, who helped in developing the model of the phenol-formaldehyde reaction.

Contents

Chapter 1
Introduction

1.1 Overview of the Main Topics

A new chemical process may involve the production of innovative chemicals, the exploitation of a new raw material, or the revamping of an established process. Irrespective of those details, the process development is usually initiated with the assessment of a new chemical route from raw materials to products, a task which requires a sound chemical skill for the understanding of the reaction mechanism, and is concluded with the assessment of the operating protocols of the industrial plant, a task which requires a sound engineering skill for obtaining a satisfactory performance of the plant, in terms of safety of operations, quality of products, and productivity.

Control and monitoring of the chemical reactor play a central role in this procedure, especially when batch operations are considered because of the intrinsic unsteady behavior and the nonlinear dynamics of the batch reactor. In order to meet such requirements, the following fundamental problems must be solved:

- **Modeling**. Mathematical modeling of an industrial plant provides the required quantitative description of the process. Mathematical models of batch reactors may include mass and energy conservation, chemical kinetics, heat exchange, and nonideal fluid dynamics; they can be used for simulation, sensitivity analysis, identification, control, and diagnosis. The development of reliable mathematical models of industrial processes and plants is often a complex and time-consuming task, which may conflict with the objective of achieving a short time-to-market strategy, so that the development of simple models, readily accessible to process engineers and sufficiently accurate, is a major challenge.
- **Identification**. In most cases, the mathematical models of interest in industry contain a few parameters whose values, essentially unknown a priori, must be computed on the basis of the available experimental data. In the case considered here, chemical kinetics is the main field in which this problem is of concern. Identification provides methods for obtaining the best estimates of those parameters and for choosing (i.e., identifying) the best mathematical model among different alternatives.

F. Caccavale et al., *Control and Monitoring of Chemical Batch Reactors*,
Advances in Industrial Control,
DOI 10.1007/978-0-85729-195-0_1,

- **Control.** Usually, the temperature inside the reactor has to be carefully controlled, in order to follow a desired profile (determined, e.g., on the basis of product/quality optimization techniques). Nevertheless, this goal is difficult to achieve, since batch reactors are often subject to large disturbances (caused by, e.g., incorrect reactor loading, fouling of internal heat exchange systems, non-ideal mixing), modeling uncertainties, incomplete real-time measurements (since chemical composition measurements are usually not available in real time), and process/equipments constraints. Since the ability of influencing its behavior decreases as the reaction proceeds, effective and industrially viable temperature control strategies have to be devised. To this aim, the use of a mathematical model of the reactor is expected to provide a significant improvement of the performance, with respect to those achieved by classical linear (e.g., PID regulators) control techniques. This motivates the focus on model-based control approaches in this book, as well as a critical comparison with more traditional linear approaches.
- **Fault diagnosis and accommodation**. Industrial plants require an high level of equipment and operational safety; such issues become critical especially in chemical industry. Hence, both equipment failures (e.g., faults affecting sensors, valves, and other devices acting on the plant) and process unexpected behaviors (e.g., temperature runaway) need to be detected in their early stages, so that corrective actions can be planned in a timely and effective way. Devising reliable and industrially viable fault diagnosis approaches is thus a major challenge. Integration of a mathematical model into the diagnosis algorithms is expected to provide major benefits in terms of both timing of the warnings and accuracy of fault identification. Hence, in this book, the focus is on model-based fault diagnosis approaches.

In the following, the reader is introduced to the book contents by illustrating in more detail the way in which the above issues are discussed throughout the book.

1.2 The Batch Reactor

The chemical batch reactor is the main object of this book and of Chap. 2, in which different aspects are considered. The chapter is opened by a classification of the ideal chemical reactors, which are simplified models of real reactors very useful as a first approach to this very complex matter. The Batch Reactor (BR) is singled out among the other ideal reactors on the basis of the mode of operation (i.e., discontinuous vs. continuous) and of the quality of mixing (i.e., perfect mixing vs. no mixing). In more general terms, a discontinuous or batch reactor corresponds to a closed thermodynamic system, whereas continuous reactors (Continuous Stirred Tank Reactor, CSTR, and Pug Flow Reactor, PFR) correspond to open systems. In industry, discontinuous operations are well suited for the production of valuable products through rather slow reactions and allow to drive reaction patterns by controlling the whole temperature–time history, whereas continuous operations in (approximatively) steady-state conditions are typical of large productions of more simple chemistry.

Chemical kinetics plays a major role in modeling the ideal chemical batch reactor; hence, a basic introduction to chemical kinetics is given in the chapter. Simplified kinetic models are often adopted to obtain analytical solutions for the time evolution of concentrations of reactants and products, while more complex kinetics can be considered if numerical solutions are allowed for.

Since complex systems may involve up to several hundreds (and even thousands) of chemical species and reactions, simple reaction pathways cannot always be recognized. In these cases, the true reaction mechanism remains an ideal matter of principle, which can be only approximated by reduced reaction networks. Also in simpler cases, reduced networks are more suitable for most practical purposes. Moreover, the relevant kinetic parameters are mostly unknown or, at best, very uncertain, so that they must be evaluated by exploiting adequate experimental campaigns. With the aim of presenting an example of the problems related to chemical kinetics, a case study is introduced and discussed in detail in the next subsection.

The mathematical model of the batch reactor consists of the equations of conservation for mass and energy. An independent mass balance can be written for each chemical component of the reacting mixture, whereas, when the potential energy stored in chemical bonds is transformed into sensible heat, very large thermal effects may be produced.

The equation of energy conservation allows one to introduce elements of realism in the modeling of the batch reactor, in particular the heat exchange apparatus. This opens the way to the arguments of thermal stability and control discussed in the second part of the book but also introduces the task of measuring and manipulating the reactor status. Hence, in the chapter a short account is given of the main measurable variables and of the main strategies for controlling the reactor temperature.

1.2.1 The Case Study

In Chaps. 2 to 6, a case study is developed in order to apply and test the methods developed along the whole book. To this purpose, the reaction between phenol and formaldehyde for the production of a prepolymer of phenolic resins has been chosen for several reasons. In fact, this reactive system is widely used in different forms for the production of different polymers; moreover, it is characterized by a noticeable production of heat and by a complex kinetic behavior. Such features represent strong challenges for controlling and monitoring tasks.

Two different classes of chemical reactions are singled out, namely the reactions of addition of formaldehyde to the aromatic ring, which introduce a methylol group as a substituent, and the reactions of condensation, which produce components with higher molecular weight. In the presence of an alkaline catalyst, the reactions of addition are strongly oriented in the *-orto* and *-para* positions of the aromatic ring, whereas the reactions of condensation occur both between two substituted positions and between a substituent and a free position, thus producing a large number of isomers.

Under suitable simplifying assumptions, a kinetic mechanism based on 13 components and 89 second-order reactions is developed. The relevant kinetic parameters (preexponential factors, activation energies, and heats of reaction) are computed on the basis of literature information. In the subsequent chapters, this kinetic model is used to test the techniques for identification, thermal stability analysis, control, and diagnosis of faults presented.

1.3 Identification of Mathematical Models

Chapter 3 provides an introduction to the identification of mathematical models for reactive systems and an extensive review of the methods for estimating the relevant adjustable parameters. The chapter is initiated with a comparison between Bayesian approach and Poppers' falsificationism. The aim is to establish a few fundamental ideas on the reliability of scientific knowledge, which is based on the comparison between alternative models and the experimental results, and is limited by the nonexhaustive nature of the available theories and by the unavoidable experimental errors.

This comparison is performed on the basis of an optimality criterion, which allows one to adapt the model to the data by changing the values of the adjustable parameters. Thus, the optimality criteria and the objective functions of maximum likelihood and of weighted least squares are derived from the concept of conditioned probability. Then, optimization techniques are discussed in the cases of both linear and nonlinear explicit models and of nonlinear implicit models, which are very often encountered in chemical kinetics. Finally, a short account of the methods of statistical analysis of the results is given.

The chapter ends with a case study. Four different reduced kinetic models are derived from the detailed kinetic model of the phenol–formaldehyde reaction presented in the previous chapter, by lumping the components and the reactions. The best estimates of the relevant kinetic parameters (preexponential factors, activation energies, and heats of reaction) are computed by comparing those models with a wide set of simulated isothermal experimental data, obtained via the detailed model. Finally, the reduced models are validated and compared by using a different set of simulated nonisothermal data.

1.4 Thermal Stability

Chapter 4 represents a bridge between Chaps. 2 and 3, which are mainly devoted to the assessment of the basic ideas of modeling and identification, and Chaps. 5 and 6, in which innovative approaches to model-based control and fault diagnosis for batch reactors are developed. In fact, this chapter discusses the thermal and chemical stability of batch reactors, thus introducing the reader to the need for adequate methods of control and fault diagnosis.

Exothermic reactions not adequately mitigated by the heat exchange system can produce very high values of the final temperature; the analysis of chemical kinetics allows us to conclude that temperature increases occur with a self-accelerating behavior, i.e., with increasing values of the relevant time derivatives. Moreover, in systems showing a more complex reaction chemistry, the increase of temperature can activate side reactions, characterized by larger values of activation energy, thus leading to a faster and, eventually, larger heat release.

In real systems, the increase of temperature is accompanied by a corresponding increase of pressure, which may lead to an explosion (i.e., to an uncontrolled increase of pressure). Nevertheless, the analysis of temperature patterns with simple kinetics is enough to study the problem for adiabatic reactors and for constant wall temperature (isoperibolic) reactors, whereas the more complex case of controlled wall temperature requires the adoption of more advanced methods.

Thus, the equations describing the thermal stability of batch reactors are written, and the relevant dimensionless groups are singled out. These equations have been used in different forms to discuss different stability criteria proposed in the literature for adiabatic and isoperibolic reactors. The Semenov criterion is valid for zero-order kinetics, i.e., under the simplifying assumption that the explosion occurs with a negligible consumption of reactants. Other classical approaches remove this simplifying assumption and are based on some geometric features of the temperature–time or temperature–concentration curves, such as the existence of points of inflection and/or of maximum, or on the parametric sensitivity of these curves.

Finally, the application of some of those criteria to the phenol–formaldehyde reaction gives some interesting insights on the thermal behavior of the system and also highlights the operation limits arising from an imposed maximum allowable temperature in the reactor.

1.5 Control of Batch Reactors

Chapter 5 is focused on the temperature control of chemical batch reactors, with special emphasis on model-based control approaches.

Control of the temperature allows one to determine the behavior of the chemical reaction and thus the final product of the batch. Of course, temperature control is of the utmost importance to ensure safety of the plant and the human operators, especially in the presence of highly exothermic reactions, where the amount of heat released can become very large, and, if the heat generated exceeds the cooling capability, temperature runaway may occur. In industrial practice the temperature can be controlled via the heat exchange between the reactor and a heating/cooling fluid, circulating in a jacket surrounding the vessel, or in a coil inside the vessel. The control approaches developed in the chapter can be adopted for different cooling/heating systems.

The chapter provides an overview of the most commonly adopted feedback control strategies, ranging from conventional linear PID controllers to more sophisticated nonlinear approaches. Since batch industrial processes can exhibit highly nonlinear behavior and operate within a wide range of conditions, linear controllers must be tuned very conservatively, in order to provide a stable behavior over the entire range of operation, thus leading to a degradation of performance. Hence, in the last two decades, nonlinear model-based control strategies began to be preferred for complex processes, thanks to the development of accurate experimental identification methods for nonlinear models and to significant improvements of computing hardware and software.

Therefore, the chapter is mainly focused on the design of model-based control approaches. Namely, a controller–observer control strategy is considered, where an observer is designed to estimate the heat released by the reaction, together with a cascade temperature control scheme. The performance of this control strategy are further improved by introducing an adaptive estimation of the heat transfer coefficient. Finally, the application of the proposed methods to the phenol–formaldehyde reaction studied in the previous chapters is presented.

1.6 Fault Diagnosis for Chemical Batch Reactors

Chapter 6 is focused on fault diagnosis methods for chemical batch processes. Consistent with the approach followed in Chap. 5, the focus of the chapter is on model-based techniques and, in particular, on techniques based on the use of state observers.

Several kinds of failures may compromise safety and productivity of industrial processes. Indeed, faults may affect the efficiency of the process (e.g., lower product quality) or, in the worst scenarios, could lead to fatal accidents (e.g., temperature runaway) with injuries to personnel, environmental pollution, and equipments damage. In the chemical process fault diagnosis area, the term *fault* is generally defined as a departure from an acceptable range of an observed variable or a parameter. Fault diagnosis (FD) consists of three main tasks: *fault detection*, i.e., the detection of the occurrence of a fault, *fault isolation*, i.e., the determination of the type and/or the location of the fault, and *fault identification*, i.e., the determination of the fault profile. After a fault has been detected, controller reconfiguration for the self-correction of the fault effects (*fault accommodation*) can be achieved in some cases.

In the chapter, first the basic principles of model-based FD are reviewed, together with a wide literature review. Then, the problem of model-based FD for chemical batch reactors is presented in detail, where both process/actuator faults (e.g., failures of the heating/cooling systems) and sensor faults (i.e., failures of the temperature sensors) are considered. In detail, a bank of two observers is designed to achieve sensors fault detection and isolation, whereas a suitable voting scheme is adopted to output an estimate of the healthy measured signals. As for process/actuator faults, a bank of observers is designed to detect, isolate, and estimate faults belonging to a finite set of fault types.

A case study, referred to the phenol–formaldehyde reaction model developed in the previous chapters, closes the chapter.

1.7 Applications to Non-ideal Reactors

This last chapter sketches the extension of the methods developed in the previous chapters to real chemical batch reactors, characterized by nonideal fluid dynamics and by the presence of multiphase systems.

First, different typologies of nonideal batch reactors are considered. In particular gas–liquid reactors are discussed, which may be used for different industrial applications (e.g., reactions of oxidation) and are often encountered in the case of gassy reactions (i.e., liquid-phase reactions which do not produce significant thermal effects but in which the production of gaseous products may lead to explosions).

The effects deriving from both nonideal mixing and the presence of multiphase systems are considered, in order to develop an adequate mathematical modeling. Computational fluid dynamics models and zone models are briefly discussed and compared to simpler approaches, based on physical models made out of a few ideal reactors conveniently connected.

The nonideal behavior also depends on reactor dimensions; thus scale-up methods are sketched, in order to face the problems deriving from the industrial scale of those reactors.

On the basis of these arguments, the chapter and the book concludes with a few suggestions for developing future research work in this field, for applying the methods presented in this book to real reactors, and for improving the proposed control and diagnosis strategies.

1.8 Suggested Reading Paths

The aim and the hope of the authors is to provide, through this book, a unitary perspective of the main problems and challenges related to modeling, control, and diagnosis of chemical batch reactors. A special emphasis is put on the interaction between the development of effective and reliable mathematical models of the plant and on the subsequent design of the control and diagnosis systems. Hence, the recommendation for the reader is to read this monograph as a whole.

However, depending on the main interests and background of the reader, two main reading paths can be identified. The first, suggested to readers mainly interested in modeling and performance evaluation issues, is composed by Chaps. 2, 3, and 4. Readers mainly interested in control and diagnosis methods are invited to read Chaps. 2, 3, 5, and 6.

Chapter 2
The Chemical Batch Reactor

List of Principal Symbols

A	reactant A
B	reactant B
c	mass heat capacity [$\text{J kg}^{-1}\,\text{K}^{-1}$]
C	concentration [mol m^{-3}]
DPh	aggregate dimers
E_a	activation energy [J mol^{-1}]
ΔE_R	internal energy change of reaction [J mol^{-1}]
F	formaldehyde
F_V	volumetric flow rate [$\text{m}^3\,\text{s}^{-1}$]
F_M	molar flow rate [mol s^{-1}]
ΔH_R	molar enthalpy change of reaction [J mol^{-1}]
I	reaction intermediate
k_0	preexponential factor [$(\text{mol m}^{-3})^{1-n}\,\text{s}^{-1}$]
k_c	rate constant [$(\text{mol m}^{-3})^{1-n}\,\text{s}^{-1}$]
K_{eq}	equilibrium constant
m	mass [kg]
MPh	mono- and di-methylolphenols
n	order of reaction
N	number of moles [mol]
N_C	number of species
N_R	number of reactions
P	reaction product
Ph	phenol
PPh	polyphenols
R	reaction rate [$\text{mol m}^{-3}\,\text{s}^{-1}$]
$\mathcal{R}$	universal gas constant [$\text{J mol}^{-1}\,\text{K}^{-1}$]
$R^\bullet$	radical species
S	heat transfer area [m^2]
$\mathcal{S}$	selectivity
t	time [s]

F. Caccavale et al., *Control and Monitoring of Chemical Batch Reactors*,
Advances in Industrial Control,
DOI 10.1007/978-0-85729-195-0_2,

t_b batch time [s]
t_P residence time [s]
T temperature [K]
TMPh trimethylolphenol
U overall heat transfer coefficient [$\mathrm{J\,m^{-2}\,K^{-1}\,s^{-1}}$]
V volume [$\mathrm{m^3}$]
X degree of conversion

Greek Symbols
ρ density [$\mathrm{kg\,m^{-3}}$]
υ stoichiometric coefficient

Subscripts and Superscripts
a ambient conditions
A reactant A
ad adiabatic conditions
B reactant B
in inlet
j jacket
max maximum
min minimum
out outlet
r reactor
0 initial value
° reference value

2.1 Ideal Chemical Reactors

Chemical reactions occur almost everywhere in the environment; however, a chemical reactor is defined as a device properly designed to let reactions occur under controlled conditions toward specified products. To a visual observation, chemical reactors may strongly differ in dimensions and structure; nevertheless, in order to derive a mathematical model for their quantitative description, essentially two major features are to be considered: the mode of operation and the quality of mixing.

Therefore, the analysis of the main object of this book, namely, the batch chemical reactor, can start by considering the different ideal chemical reactors. In fact, ideal reactors are strongly simplified models of real chemical reactors [10], which however capture the two major features mentioned above. These models can be classified according to the mode of operation (i.e., discontinuous vs. continuous) and to the quality of mixing (i.e., perfect mixing vs. no mixing). The three resulting ideal reactors are sketched in Fig. 2.1.

The discontinuous stirred reactor (Batch Reactor, BR, Fig. 2.1(a)) corresponds to a closed thermodynamic system, whereas the two continuous reactors (Continuous Stirred Tank Reactor, CSTR, Fig. 2.1(b), and Plug Flow Reactor, PFR, Fig. 2.1(c))

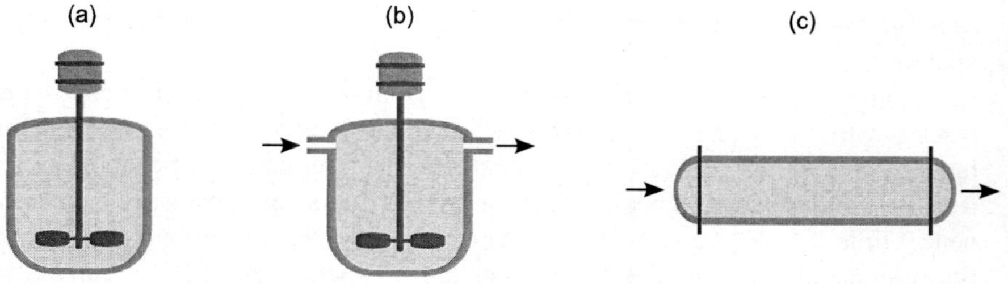

Fig. 2.1 Ideal reactors: BR (**a**), CSTR (**b**), and PFR (**c**)

are open systems. In industry, discontinuous operations are well suited for the production of valuable products through rather complex reactions and allow one to drive the reaction pattern by controlling the temperature, whereas continuous operations in (approximately) steady-state conditions are typical of large productions, usually based on a more simple chemistry.

The two extreme hypotheses on mixing produce lumped models for the fluid dynamic behavior, whereas real reactors show complex mixing patterns and thus gradients of composition and temperature. It is worthwhile to stress that the fluid dynamic behavior of real reactors strongly depends on their physical dimensions. Moreover, in ideal reactors the chemical reactions are supposed to occur in a single phase (gaseous or liquid), whereas real reactors are often multiphase systems. Two simple examples are the gas–liquid reactors, used to oxidize a reactant dissolved in a liquid solvent and the fermenters, where reactions take place within a solid biomass dispersed in a liquid phase. Real batch reactors are briefly discussed in Chap. 7, in the context of suggestions for future research work.

Those simplified models are often used together with simplified overall reaction rate expressions, in order to obtain analytical solutions for concentrations of reactants and products. However, it is possible to include more complex reaction kinetics if numerical solutions are allowed for. At the same time, it is possible to assume that the temperature is controlled by means of a properly designed device; thus, not only adiabatic but isothermal or nonisothermal operations as well can be assumed and analyzed.

The main ideas of chemical kinetics are reviewed in the next section; for the sake of completeness, a brief account is given here of the performance of continuous reactors as compared to BR, which is the object of the present book.

Whereas the operation of batch reactors is intrinsically unsteady, the continuous reactors, as any open system, allow for at least one reacting steady-state. Thus, the control problem consists in approaching the design steady-state with a proper startup procedure and in maintaining it, irrespective of the unavoidable changes in the operating conditions (typically, flow rate and composition of the feed streams) and/or of the possible failures of the control devices. When the reaction scheme is complex enough, the continuous reactors behave as a nonlinear dynamic system and show a complex dynamic behavior. In particular, the steady-state operation can be hindered by limit cycles, which can result in a marked decrease of the reactor performance. The analysis of the above problem is outside the purpose of the present text;

nevertheless, a few interesting observations can be made on the simple steady-state operation.

Apparently, the PFR differs more strongly from the BR, since it is a continuous reactor with no mixing. Nevertheless, when the PFR is described in the Eulerian mode, it appears as made of infinitesimal reaction volumes, dV, behaving as differential batch reactors, since they remain in the reactor for a residence (or permanence) time $t_P = V_r/F_V$ (where V_r is the reactor volume, and F_V is the volumetric flow rate passing through the reactor) and do not experience relative mixing. Thus, this reactor can be described by the same equations of the batch reactor, when t_P is considered in lieu of the time variable t. It is worth remarking that, for any fixed reactor volume, t_P can be changed by changing F_V, e.g., in order to optimize the reactor performance.

For the perfectly mixed continuous reactor, the CSTR, the ratio V_r/F_V only represents the mean residence time, $t_{P,av}$; however, it is still possible to compare the performance of the CSTR with the performance of the BR by letting the mean residence time $t_{P,av} = t$. Interestingly, when the reaction rate shows a positive dependence on reactants concentration, the BR is more effective than the CSTR. This is because the batch reactor experiences all the system compositions between initial and final values, whereas the CSTR operates at the final composition, where the reaction rate is smaller (under the above hypotheses). Finally, one can compare the two continuous reactors under steady-state conditions. The CSTR allows a more stable operation because of back-mixing, which however reduces the chemical performance, whereas the PFR is suitable for large heat transfer but suffers from larger friction losses.

2.2 The Rate of Chemical Reactions

Chemical reactions change the molecular structure of matter, thus resulting in the destruction of some chemical species (reactants) and in the formation of different ones (products). The relevant quantities of reactants and products involved in the reaction are strictly determined by stoichiometry, which states a law of proportionality deriving from the mass conservation of the single elements. Often, the stoichiometric coefficients are imposed to be constant during the reaction; however, this is not true in most real systems. When variable stoichiometric coefficients are observed, the system cannot be described by a single reaction.

With reference to a simple reaction with constant stoichiometric coefficients, and unless otherwise specified, the reaction rate R [moles time^{-1} volume^{-1}] measures the specific velocity of destruction of those reactants (and of formation of those products) that appear with unitary stoichiometric coefficients. The reaction rates of each other component are proportional to R according to their stoichiometric coefficients.

In general, the rate of a chemical reaction can be expressed as a function of chemical composition and temperature. This function usually takes the form of a power law with respect to reactant concentrations and of an exponential function in the inverse absolute temperature. As an example, the rate R of conversion of A and

B in the reaction

$$A + B \rightarrow \text{products} \tag{2.1}$$

can be expressed as

$$R = k_c(T_r) C_A^{n_A} C_B^{n_B} = k_0 \exp\left(-\frac{E_a}{\mathcal{R} T_r}\right) C_A^{n_A} C_B^{n_B}, \tag{2.2}$$

where C_A and C_B are the molar concentrations of reactants, n_A and n_B are the orders of reaction ($n = n_A + n_B$ being the overall reaction order), $k_c(T_r)$ is the rate constant, k_0 is the preexponential factor, E_a is the activation energy, $\mathcal{R}$ is the universal gas constant, and T_r is the absolute reaction temperature. Since, on varying temperature from 0 to ∞, the S-shaped function $\exp(-E_a/\mathcal{R}T_r)$, known as Arrhenius law or Arrhenius term, ranges from 0 to 1, the preexponential factor k_0 represents the limit of k_c as $T_r \rightarrow \infty$.

Function (2.2) can be considered as an empirical model used to best fit the experimental concentration-time data. In practice, laws different from (2.2) are also encountered, especially when dependence on the concentration is considered; however, a simple theory based on the kinetic theory of gases can only explain the simplest of these empirical rate laws. The general idea of this theory is that reaction occurs as a consequence of a collision between adequately energized molecules of reactants. The frequency of collision of two molecules can explain simple reaction orders, namely the schemes

$$A + B \rightarrow \text{products} \quad (R_A \propto C_A C_B), \tag{2.3}$$

$$A + A \rightarrow \text{products} \quad \left(R_A \propto C_A^2\right), \tag{2.4}$$

$$A + \text{third body} \rightarrow \text{products} \quad (R_A \propto C_A), \tag{2.5}$$

where *third body* stands for any molecule with constant concentration. Any collision involving more than two molecules is very unlikely and must be neglected.

On the other hand, the effective collision concept can explain the Arrhenius term on the basis of the fraction of molecules having sufficient kinetic energy to destroy one or more chemical bonds of the reactant. More accurately, the formation of an *activated complex* (i.e., of an unstable reaction intermediate that rapidly degrades to products) can be assumed. Theoretical expressions are available to compute the rate of reaction from thermodynamic properties of the activated complex; nevertheless, these expression are of no practical use because the detailed structure of the activated complexes is unknown in most cases. Thus, in general the kinetic parameters (rate constants, activation energies, orders of reaction) must be considered as unknown parameters, whose values must be adjusted on the basis of the experimental data.

Chemical reactions occurring because of a single kinetic act, i.e., because of a single collision between two molecules, are defined as *elementary reactions*. More complex laws of dependence on concentrations can be explained by complex reaction mechanisms, i.e., by the idea that most reactions occur as a sequence of many elementary reactions, linked in series or in parallel. As an example, the following simple reaction mechanism, made out of two reaction steps in series, can explain a fractionary reaction order. Let us consider the reaction

$$A + B \rightarrow P, \tag{2.6}$$

for which a first-order behavior with respect to each reactant can be foreseen on the basis of the collision theory. Nevertheless, suppose that this reaction is not caused by a single collision, but with the following mechanism: first, reactant A is in equilibrium with a reaction intermediate I since both direct and inverse reactions are very fast:

$$A \rightleftharpoons 2I, \tag{2.7}$$

then, I reacts with B producing P,

$$I + B \rightarrow P. \tag{2.8}$$

By applying the result (2.3) to reaction (2.8) and introducing the equilibrium constant, K_{eq}, for the reaction (2.7), defined as

$$K_{eq} = \frac{C_I^2}{C_A}, \tag{2.9}$$

one obtains

$$R = k_c C_I C_B = k_c (K_{eq} C_A)^{1/2} C_B. \tag{2.10}$$

The apparent rate constant in (2.10), which is obtained by multiplying a true rate constant k_c and the square root of an equilibrium constant, K_{eq}, can show a law of dependence on temperature different from the simple Arrhenius law. In some cases, even a negative temperature dependence can be observed. Moreover, if both mechanisms (2.6) and (2.7)–(2.8) are active in parallel, the observed reaction rate is the sum of the single rates, and an effective reaction order variable from $1/2$ to 1 can be observed with respect to reactant A. Variable and fractionary reaction orders can be also encountered in heterogeneous catalytic reactions as a consequence of the adsorption on a solid surface [6].

Very fast reactions, such as combustion reactions, are very often characterized by chain mechanisms activated by very reactive species, such as radicals. First, radicals, $R_1^\bullet$, are formed by an opening reaction involving the reactant A,

$$A \longrightarrow 2R_1^\bullet; \tag{2.11}$$

then, the chain is propagated by a loop of reactions that continuously produces the final product P and regenerates the radicals:

$$R_1^\bullet + B \longrightarrow P + R_2^\bullet, \tag{2.12}$$

$$R_2^\bullet + A \longrightarrow P + R_1^\bullet. \tag{2.13}$$

Moreover, branching reaction mechanisms can take place when at least one reaction leads to multiplication of radicals, such as

$$R_1^{\bullet} + A \longrightarrow P + R_2^{\bullet} + R_3^{\bullet}. \tag{2.14}$$

In this case, the fast increase of concentration of radicalic species can result in the loss of control of the reaction (runaway) and in the explosion of the system. This radicalic runaway may be strongly enhanced by linked thermal effects that are discussed in more details in Chap. 4.

Kinetic mechanisms involving multiple reactions are by far more frequently encountered than single reactions. In the simplest cases, this leads to reaction schemes in series (at least one component acts as a reactant in one reaction and as a product in another, as in (2.7)–(2.8)), or in parallel (at least one component acts as a reactant or as a product in more than one reaction), or to a combination series-parallel. More complex systems can have up to hundreds or even thousands of intermediates and possible reactions, as in the case of biological processes [12], or of free-radical reactions (combustion [16], polymerization [4]), and simple reaction pathways cannot always be recognized. In these cases, the true reaction mechanism mostly remains an ideal matter of principle that can be only approximated by reduced kinetic models. Moreover, the values of the relevant kinetic parameters are mostly unknown or, at best, very uncertain.

The model reduction procedure must be adapted to the use of the simplified models and to the availability of experimental data needed to evaluate the unknown parameters, as discussed in Chap. 3. In general, more complex models are used for the design of the reactor and for the simulation of the entire process, whereas more simplified models are best fit for feedback control. In the following chapters it is shown that fairly accurate results are obtained when a strongly simplified kinetic model is used for control and fault diagnosis purposes.

2.3 The Ideal Batch Reactor

A more quantitative analysis of the batch reactor is obtained by means of mathematical modeling. The mathematical model of the ideal batch reactor consists of mass and energy balances, which provide a set of ordinary differential equations that, in most cases, have to be solved numerically. Analytical integration is, however, still possible in isothermal systems and with reference to simple reaction schemes and rate expressions, so that some general assessments of the reactor behavior can be formulated when basic kinetic schemes are considered. This is the case of the discussion in the coming Sect. 2.3.1, whereas nonisothermal operations and energy balances are addressed in Sect. 2.3.2.

2.3.1 Conservation of Mass

An independent mass balance can be written for each chemical species (or component of the reacting system) in the reactor. Let $N_i = V_r C_i$ denote the molar quantity of the ith species, where V_r is the volume of the reactor. Assuming a single reaction with rate R, the rate of change of the molar quantity, $\dot{N}_i = dN_i/dt$ [moles time^{-1}], must be equal to the rate of reaction taken with the proper algebraic sign, i.e.,

$$\dot{N}_i = \upsilon_i R V_r, \tag{2.15}$$

where υ_i is the stoichiometric coefficient of the ith component, taken negative if this component is a reactant and positive if it is a product. Since the reaction rate is a function of concentrations, it is useful to explicate the accumulation term as

$$\dot{N}_i = V_r \dot{C}_i + C_i \dot{V}_r, \tag{2.16}$$

which, under the assumption of constant volume of reaction, gives

$$\dot{C}_i = \upsilon_i R. \tag{2.17}$$

It appears that, in the case of constant volume BR, the reaction rate is strictly linked to the time derivatives of concentrations. This result, which cannot be generalized to different reactors, may be however useful to visualize the concept of reaction rate.

When multiple reactions occur simultaneously, the right-hand side of (2.17) is replaced by a sum of reaction terms

$$\dot{C}_i = \sum_{j=1}^{N_R} \upsilon_{i,j} R_j, \tag{2.18}$$

where N_R is the total number of reactions and $\upsilon_{i,j}$ is the stoichiometric coefficient of component i in reaction j, again taken negative if component i is a reactant in reaction j, positive if it is a product, and null if it is not involved. Hence, if N_C species are involved in the reaction, a set of N_C equations in the form (2.18) can be written, eventually in compact matrix form.

Table 2.1 reports some of the most classical basic reaction schemes encountered in chemical engineering, together with the explicit expressions of the isothermal concentration profiles as functions of time. The effect of the reaction order can be evaluated by considering the first three cases in Table 2.1; by applying the corresponding rate laws, the curves shown in Fig. 2.2 are obtained. To allow an easier comparison, the values of the rate constants have been chosen so as to obtain the same C_A at an arbitrary batch time t_b.

The zero-order kinetics is characterized by a linear concentration profile, which is however unrealistic at very large reaction times, since it produces a negative reactant concentration; this result confirms that a zero-order reaction derives from a complex reaction mechanism that cannot be active at very low reactant concentrations. On increasing the reaction order, the reaction is faster at the highest concentration values

Table 2.1 Simple reaction schemes

No.		Kinetic scheme	Integrated BR model equation
1	zero order	$A \xrightarrow{k_{c0}} P$	$C_A = C_{A0} - k_{c0}t$
2	first order	$A \xrightarrow{k_{c1}} P$	$C_A = C_{A0}\exp(-k_{c1}t)$
3	second order	$A \xrightarrow{k_{c2}} P$	$C_A = \dfrac{C_{A0}}{1 + k_{c2}t}$
4	equilibrium	$A \underset{k_{cA}}{\overset{k_{cB}}{\rightleftharpoons}} B$	$C_A = \frac{C_{A0}}{1+K_{eq}}[1 - \exp(-(k_{cA} + k_{cB})t)]$
5	parallel	$A \xrightarrow{k_{cP1}} P_1$	$C_A = C_{A0}\exp(-(k_{cP1} + k_{cP2})t)$
		$A \xrightarrow{k_{cP2}} P_2$	$C_{P1} = C_{A0}\frac{k_{cP1}}{k_{cP1}+k_{cP2}}[1 - \exp(-(k_{cP1} + k_{cP2})t)]$
6	series	$A \xrightarrow{k_{cI}} I \xrightarrow{k_{cP}} P$	$C_A = C_{A0}\exp(-k_{cI}t)$
			$C_I = C_{A0}\frac{k_{cI}}{k_{cP}-k_{cI}}[\exp(-k_{cI}t) - \exp(-k_{cP}t)]$
7	multiple series	$A \rightarrow \cdots \rightarrow P$	
8	series-parallel	$A \xrightarrow{k_{cI}} I \xrightarrow{k_{cP}} P$	$C_A = C_{A0}\exp(-(k_{cI} + k_{cS})t)$
		$A \xrightarrow{k_{cS}} P$	$C_I = C_{A0}\frac{k_{cI}}{k_{cP}-k_{cS}-k_{cI}}[\exp(-(k_{cI} + k_{cS})t) - \exp(-k_{cP}t)]$

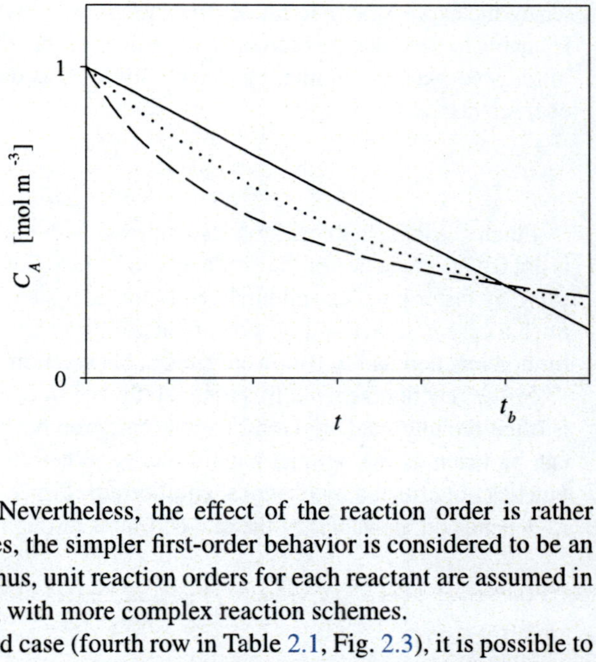

Fig. 2.2 Time histories of C_A in a batch reactor for zero (*continuous line*), first (*dotted line*) and second (*dashed line*) order reaction rates and $C_{A0} = 1\ \mathrm{mol\,m^{-3}}$

and slower at the lowest. Nevertheless, the effect of the reaction order is rather small, so that, in many cases, the simpler first-order behavior is considered to be an adequate approximation. Thus, unit reaction orders for each reactant are assumed in the following when dealing with more complex reaction schemes.

In the equilibrium limited case (fourth row in Table 2.1, Fig. 2.3), it is possible to simulate the constant C_B/C_A ratio imposed by thermodynamics by introducing the inverse reaction $B \rightarrow A$. In this case, the reaction is not complete, and an asymptotic behavior is observed for both reactant and product.

In the parallel reaction scheme (fifth row in Table 2.1), competition is observed between the two reactions when only one of the products is required and the other one is a secondary undesired or a low value product. In this case, the degree of

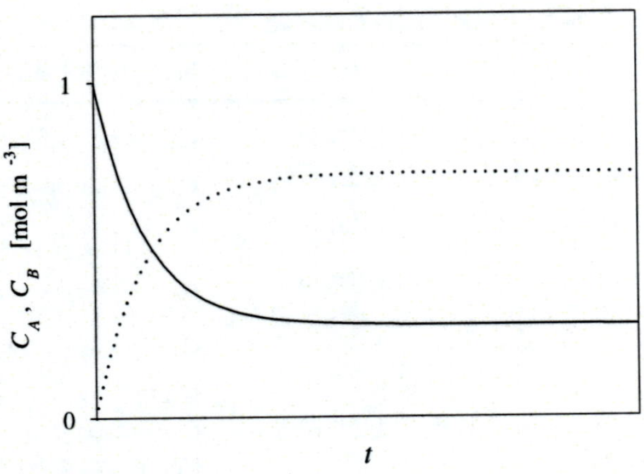

Fig. 2.3 Time histories of C_A (*continuous line*) and C_B (*dotted line*) in a batch reactor for the equilibrium limited reaction. Initial conditions are: $C_{A0} = 1$ mol m^{-3} and $C_{B0} = 0$ mol m^{-3}

conversion of the reactant, defined as

$$X = \frac{N_{A0} - N_A}{N_{A0}} = \frac{C_{A0} - C_A}{C_{A0}}, \tag{2.19}$$

where the expression in terms of concentrations holds for constant-volume reactors, is unable to describe the product distribution, so that the selectivity concept must be introduced. As an example, the selectivity to P_1 is defined, for unit stoichiometric coefficients, as

$$S_{P_1} = \frac{C_{P_1}}{C_{A0} - C_A}. \tag{2.20}$$

Finally, when chemical kinetics contrasts with equilibrium, the parallel scheme is not trivial, since one of the products can be favored in the early stages of the batch cycle by faster kinetics and hindered in the later stages by unfavorable equilibrium. Such a case is shown in Fig. 2.4 for parallel reactions of A to P_1 via an equilibrium limited reaction and to P_2 via an irreversible reaction.

In the reaction scheme in series (sixth row in Table 2.1), the required product is often the intermediate I, and its concentration has a maximum at time t^*, which can be taken as the optimal batch time, t_b. When the system follows a first-order kinetics not affected by chemical equilibrium (Fig. 2.5), it can be easily shown that t^* depends on the values of the rate constants through the following expression:

$$t^* = \frac{\ln(k_{cP}/k_{cI})}{k_{cP} - k_{cI}}. \tag{2.21}$$

It is worth remarking that, in real cases, the simple criterion $t_b = t^*$, based on the system's selectivity to the desired product, must be modified to account for the cost of operation (including separation between products and unreacted reactants) and the gross added value related to the transformation of reactants into products.

It is also interesting to note that the concentration–time curve of the final product P has a typical shape with zero derivative at $t = 0$ and an asymptotic trend at very

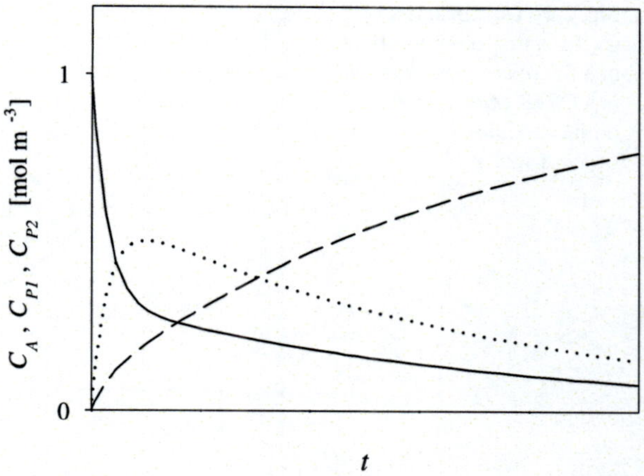

Fig. 2.4 Time histories of C_A (*continuous line*), C_{P1} (*dotted line*), and C_{P2} (*dashed line*) in a batch reactor for parallel reactions of A producing P_1, via an equilibrium limited reaction, and P_2, via an irreversible reaction. Initial conditions are: $C_{A0} = 1$ mol m^{-3}, $C_{P10} = C_{P20} = 0$ mol m^{-3}

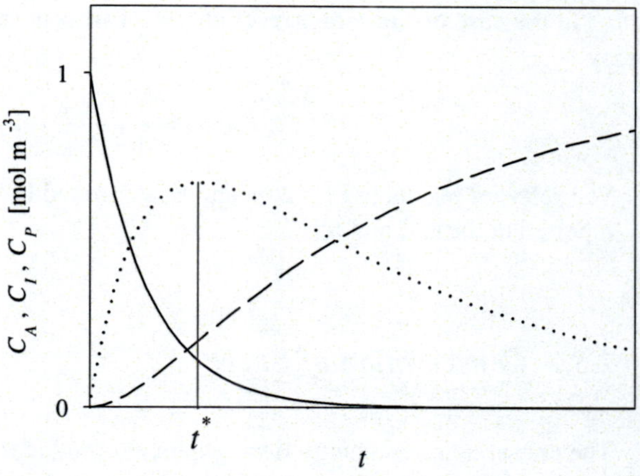

Fig. 2.5 Time histories of C_A (*continuous line*), C_I (*dotted line*), and C_P (*dashed line*) in a batch reactor for irreversible series reactions. Initial conditions are: $C_{A0} = 1$ mol m^{-3}, $C_{I0} = C_{P0} = 0$ mol m^{-3}

large times. These features are also encountered in more complex series schemes, i.e., when more than one intermediate is observed (seventh row in Table 2.1), and/or when kinetics is hindered by unfavorable equilibrium. In general, it appears that the time t^* must be considered only as a first approximation of the optimal batch time, which is computed as before on the basis of a cost analysis.

Finally, the eighth reaction mechanism in Table 2.1 includes both series and parallel reactions to the same product P. This scheme is more complete and somewhat more realistic, but it is not so much different from the series scheme, because the side parallel reaction to P only produces small changes in the shape of the concentration profiles. As an example, the initial zero derivative for C_P can be canceled.

It is also interesting to quantitatively compare the performance of a BR with those obtained by a CSTR, for which the reaction term RV_r acts as a selective stream entering or leaving the reactor; hence, the mass balance for a CSTR reads

$$F_{MA,in} = F_{V,in}C_{A,in} = F_{MA,out} + RV_r = F_{V,out}C_{A,out} + RV_r, \tag{2.22}$$

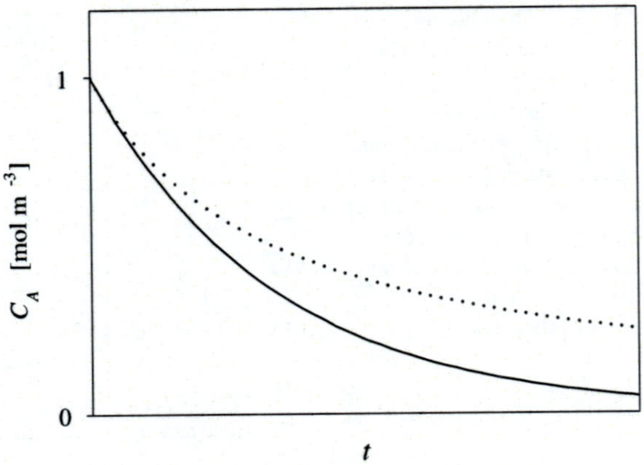

Fig. 2.6 Time histories of C_A for a first-order reaction in a BR (*continuous line*) and in a CSTR (*dotted line*). Initial condition is $C_{A0} = 1\ \text{mol}\,\text{m}^{-3}$

where $F_{MA,in}$ and $F_{MA,out}$ are, respectively, the inlet and outlet molar flow rates.

In the case of first-order reactions, the exit concentration of reactant A is given by

$$C_{A,out} = \frac{C_{A,in}}{1 + t_{P,av}k_c}. \tag{2.23}$$

The relevant results in Fig. 2.6 can be interpreted in the light of the considerations reported at the end of Sect. 2.1.

2.3.2 *Conservation of Energy*

The conservation of energy (heat balance) introduces an important element of realism into the model, i.e., the coupling of the reactor with the heating/cooling device. When the potential energy stored in chemical bonds is transformed by an exothermal chemical reaction into sensible heat, considerable thermal effects may be produced that can be quantitatively described by a proper form of the equation of energy conservation. In a batch reactor, the accumulation of internal energy is given by the difference between the heat produced by reaction and the heat exchanged with the surroundings:

$$\text{Stored Energy} = \text{Generated Heat} - \text{Exchanged Heat}. \tag{2.24}$$

A few simplified assumptions make this equation of practical utility. The left-hand side in (2.24), i.e., the rate of change of internal energy [energy time^{-1}], is simply related to the total mass m of reaction solution, to the overall constant volume specific heat capacity c_{vr} [energy mass^{-1} temperature^{-1}], and to the rate of change of reactor temperature $\dot{T}_r$. The heat generated by chemical reaction is given by the product of the specific molar energy change due to reaction, ΔE_R, and the amount of moles converted in the reactor per unit time, RV_r.

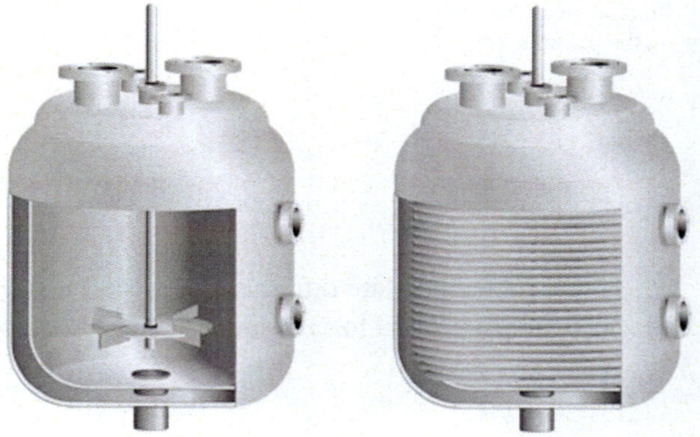

Fig. 2.7 Batch reactor with external heat exchange jacket (*left*) and coil (*right*)

The values of ΔE_R can be computed from the standard internal energy change ΔE_R^o, which refers to reactants and products in their standard states (not mixed, at 1 atm and 25°C) but also depends on temperature and, for nonideal solutions, on the heat of mixing of the components. Since a detailed description of these second-order thermal effects is beyond the purposes of a standard modeling approach, this quantity can be approximated by the standard molar enthalpy change (usually named standard heat of reaction), ΔH_R^o, which can be easily computed from available tables of standard enthalpy of formation of the individual compounds. Since ΔH_R^o is positive for endothermic reactions, a minus sign is usually introduced in the energy balance. Consistent with this simplified assumption, in liquid-phase systems the (very small) difference between the constant-pressure and constant-volume heat capacities can be neglected; hence, the heat capacity is hereafter denoted by c_r, without any further specification.

The second term on the right-hand side of (2.24) depends on the modes of heat exchange between the reactor and a heat exchange medium or the surroundings. In general, in order to accomplish the different stages of a batch operation (initial reactor heating, reaction development, and final cooling), the reactor must be provided with a properly designed device for heat exchange. A jacket or a coil, as depicted in Fig. 2.7, are suitable for heating (e.g., by using hot water or steam) and cooling (e.g., by using cold water) only for relatively small heat loads, since the exchange area is limited by the external reactor surface.

For larger heat loads, i.e., when ΔE_R and/or R and/or V_r increase, a larger heat exchange surface must be provided. A heat exchanger made out of several tubes located inside the reactor allows one to obtain a larger surface-to-volume ratio; however, its dimensions are limited by the reactor volume and by effectiveness of mixing of the reaction media. Thus, for large heat loads, an external shell and tube heat exchanger must be designed, whose dimensions do not depend on the reactor dimensions. The reaction solution circulates from the reactor to the exchanger and then back to the reactor in a closed loop; this circulating flow also produces a positive effect on the mixing of the reactor contents.

According to Newton's law of heat exchange, the heat exchanged by the reactor depends on the overall coefficient of heat exchange, U, on the heat exchange surface

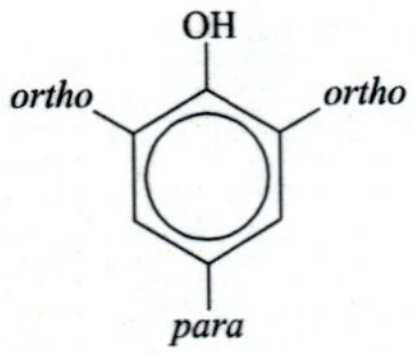

Fig. 2.8 Reactive positions on the phenolic ring

S, and on the temperature difference between the reactor and the coolant, $T_r - T_j$. In conclusion, a general form of the heat balance is given by

$$mc_r\dot{T}_r = (-\Delta H_R^o)RV_r - US(T_r - T_j). \tag{2.25}$$

For nonjacketed reactors, a further term $-U_aS(T_r - T_a)$ can be eventually added to the energy balance to account for heat losses toward reactor surroundings. Here, U_a is the overall coefficient of heat exchange with the external environment, and T_a is the external environment temperature.

When (2.25) is integrated from the initial condition $t = 0$ and $C_A = C_{A0}$ to $t \to \infty$ and $C_A \to 0$ in the case of adiabatic reactor ($US = 0$), the adiabatic temperature rise $\Delta T_{ad} = T_{ad} - T_0$ is obtained, which represents a useful measure of practical utility of the system reactivity in terms of the maximum temperature obtainable when chemical energy is entirely transformed into sensible heat.

2.4 Introducing the Case Study

In this section, the phenol–formaldehyde reaction is introduced as a case study. This reaction has been chosen because of its kinetic complexity and its high exothermicity, which poses a strong challenge for modeling and control practice. The kinetic model presented here is adopted to simulate a realistic batch chemical process; the identification, control, and diagnosis approaches developed in the next chapters are validated by resorting to this model.

Phenol (C_6H_5OH) and formaldehyde (CH_2O) can react in different ways depending on the catalyst used and the initial formaldehyde-to-phenol molar ratio. In the production of resol-type phenolic resins, in the presence of an alkaline catalyst, the reactions occurring in the system can be classified into two main types, namely, reactions of addition of methylol groups to the aromatic ring and reactions of condensation of aromatic rings to form higher molecular weight components and, finally, polymers [14].

In the alkaline solution, phenol is essentially present as phenate ion, so that the first steps of the reaction may be depicted as electrophilic additions of formaldehyde to the aromatic ring. Those attacks are essentially favored in the *-ortho* (o) and *-para* (p) positions, as sketched in Fig. 2.8, because relatively stable low activation energy intermediates can be formed; on the contrary, the attacks in the *-meta* positions are much slower and are not considered here.

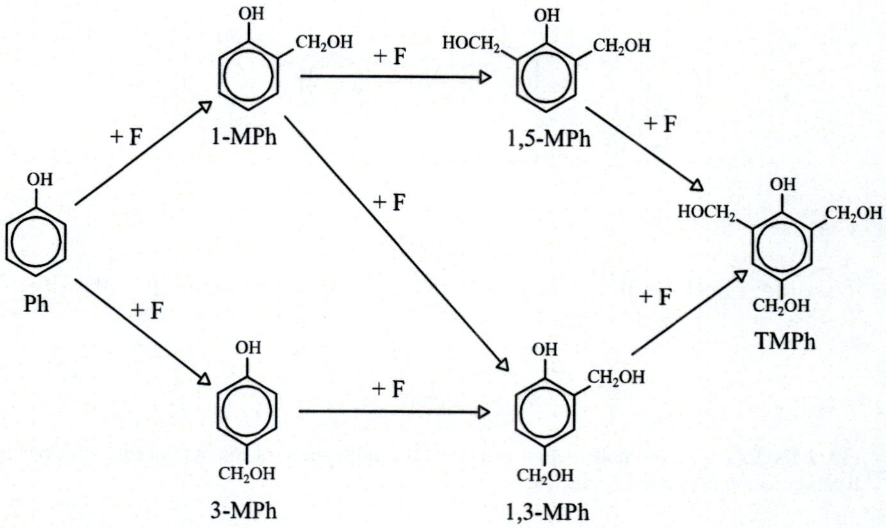

Fig. 2.9 Reactions of addition to the phenolic ring (1-MPh = o-methylolphenol, 3-MPh = p-methylolphenol, 1,5-MPh = 1,5-methylolphenol, 1,3-MPh = 1,3-methylolphenol, TMPh = 1,3,5-methylolphenol)

These reactions largely prevail in the early stages of the reaction, when both phenol and formaldehyde are present in large concentrations. This stage is characterized by the 7 reactions sketched in Fig. 2.9 and is faster than the following condensation stage; therefore, a mixture of phenol (Ph), formaldehyde (F), two mono-methylolphenols (o-methylolphenol = 1-MPh, p-methylolphenol = 3-MPh), two di-methylolphenols (1,5-methylolphenol = 1,5-MPh, 1,3-methylolphenol = 1,3-MPh), and 1,3,5-methylolphenol = trimethylolphenol = TMPh is initially formed.

The condensation reactions become important in a later stage when the concentration of the substituted phenols has increased. These reactions can occur in two different modes, as in the two examples sketched in Fig. 2.10; namely, the substituted methylol groups can react either with a nonsubstituted (free) position of a different aromatic ring (a) or with a second methylol group (b).

In the first case, the reaction produces a methylene-diphenol (MDPh), i.e., molecules in which two phenolic rings are linked by a methylene group. In the second case, the reaction produces an aromatic ether, i.e., a molecule in which two aromatic rings are linked by a dimethyl-ether group. However, these compounds are relatively unstable and rapidly decay (producing a formaldehyde molecule) to the corresponding methylene-diphenols (MDPh). Thus, the first reaction step is the rate limiting step, whereas the second one determines the final product and the total stoichiometry of the reaction.

In these reactions, a large number of two-ring isomers are formed, depending on the position of attack. In a later stage of reaction, these diphenols can undergo both addition reactions of one more methylol group on a free position of the aromatic

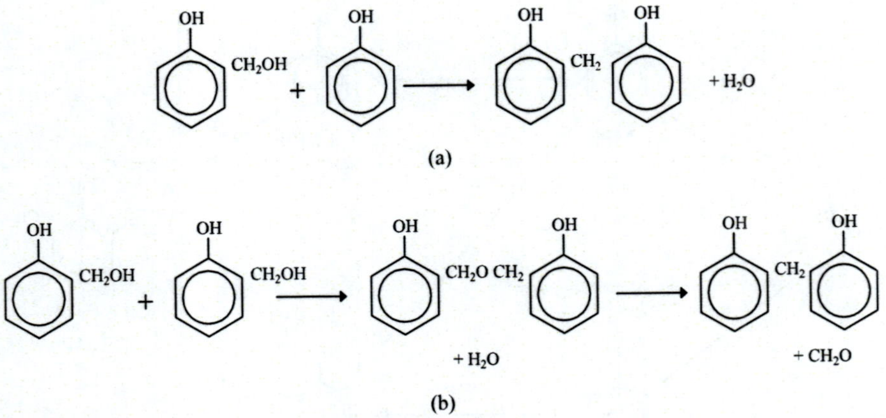

Fig. 2.10 Examples of condensation reactions: **a** condensation of methylolphenol with phenol; **b** condensation of two methylolphenols

ring and reactions of condensation to three- and four-ring molecules and finally to larger molecular weight polymers. Nevertheless, the attention is focused here on the production of TMPh, a product that has a commercial value as a prepolymer for resol-type resins; thus, the condensation reactions of higher molecular weight molecules are usually limited by quenching the system at a suitable reaction time. On the basis of the above analysis, a quantitative kinetic model is derived, as discussed in the next two sections.

2.4.1 Components

Only a few studies have tackled the problem of deriving a detailed kinetic model of the phenol–formaldehyde reactive system, mainly because of its complexity. In recent years, a generalized procedure has been reported in [11, 14] that allows one to build a detailed model for the synthesis of resol-type phenolic resins. This procedure is based on a group contribution method and virtually allows one to estimate the kinetic parameters of every possible reaction taking place in the system.

In order to develop an exhaustive kinetic model, some basic assumptions have been made. First, the two mono-methylolphenols and the two di-methylolphenols have been indicated as MPh_i ($MPh_i, i = 1, \ldots, 4$). Moreover, since the TMPh is the desired product, some simplifications have been introduced for describing the components containing more than one phenolic ring. In detail, all the diphenols having the same number of methylol groups have been considered as one single component, regardless of the position of these methylol groups. This allows one to consider in the model only 5 different aggregate methylene diphenols DPh_i ($i = 0, \ldots, 4$), where i is the number of methylol groups. In the same way, triphenols have been taken into account as just one aggregate poliphenol (PPh), whereas compounds with more than three phenolic rings have not been considered here, since the system is usually quenched before their formation becomes detectable. Therefore, the resulting reactive scheme involves 13 chemical compounds:

- reactants: phenol (Ph) and formaldehyde (F)
- intermediates: four mono- and di-methylolphenols (MPh_i, $i = 1, \dots, 4$)
- desired product: trimethylolphenol (TMPh)
- undesired products: five aggregate dimers (DPh_i, $i = 0, \dots, 4$) and
- undesired product: aggregate component PPh, including all the poliphenols with three phenolic rings.

2.4.2 Reactions

A second-order kinetics has been assumed [8, 14] for all reactions included in the reaction mechanism, which is composed by the following 89 reactions:

- 7 addition reactions of formaldehyde to phenol, sketched in Fig. 2.9
- 77 condensation reactions involving two monophenolic molecules as reactants (Ph or MPh_i, $i = 1, \dots, 4$)
- 4 addition reactions of formaldehyde to aggregate dimers; and
- 1 condensation reaction of dimers with monomers.

In order to estimate the kinetic parameters for the addition and condensation reactions, the procedure proposed in [11, 14] has been used, where the rate constant k_c of each reaction at a fixed temperature of 80°C is computed by referring it to the rate constant k_c^o at 80°C of a reference reaction, experimentally obtained. The ratio k_c/k_c^o, assumed to be temperature independent, can be computed by applying suitable correction coefficients, which take into account the different reactivity of the *-ortho* and *-para* positions of the phenol ring, the different reactivity due to the presence or absence of methylol groups and a frequency factor. In detail, the values in [11] for the resin RT84, obtained in the presence of an alkaline catalyst and with an initial molar ratio phenol/formaldehyde of $1:1.8$, have been adopted. Once the rate constants at 80°C and the activation energies are known, it is possible to compute the preexponential factors k_0 of each reaction using the Arrhenius law (2.2).

For the molar enthalpy change of reaction, the values $\Delta H_R^o = -20.3\ \text{kJ mol}^{-1}$ and $\Delta H_R^o = -98.7\ \text{kJ mol}^{-1}$ have been used for addition and condensation reactions, respectively [9].

In the following, all the reactions included in the model are reported together with the values of the relevant kinetic parameters. Addition reactions, from 1 to 7, are reported in Table 2.2, whereas condensation reactions to the single dimers (DPh_i) are reported in Tables 2.3, 2.4, 2.5, 2.6, and 2.7; for all condensation reactions, an activation energy of 90 kJ mol^{-1} has been assumed.

It should be observed that in Table 2.3 the three nonsubstituted two-ring isomers are indicated as 1,6-MDPh, 1,8-MDPh, and 3,8-MDPh, where the numbers indicate the position of the two hydroxyl groups with respect to the methylene bridge, respectively, in the o–o, o–p, and p–p positions. Those numbers are preserved in the

Table 2.2 Addition reactions producing monomers

No.	Reaction	k_0 [m^3 mol^{-1} s^{-1}]	E_a [kJ mol^{-1}]
1	Ph + F → 1-MPh	$1.13 \cdot 10^5$	89.1
2	Ph + F → 3-MPh	$2.26 \cdot 10^5$	91.7
3	o-MPh + F → 1,5-MPh	$4.34 \cdot 10^5$	98.5
4	o-MPh + F → 1,3-MPh	$2.09 \cdot 10^5$	88.2
5	p-MPh + F → 1,3-MPh	$1.11 \cdot 10^7$	99.0
6	1,5-MPh + F → 1,3,5-MPh	$1.68 \cdot 10^2$	91.5
7	1,3-MPh + F → 1,3,5-MPh	$6.99 \cdot 10^5$	92.2

Table 2.3 Condensation reactions producing the dimers DPh_0

No.	Reaction	k_0 [m^3 mol^{-1} s^{-1}]
8	Ph + 1-MPh → 1,6-MDPh	$3.65 \cdot 10^3$
9	Ph + 1-MPh → 1,8-MDPh	$3.01 \cdot 10^3$
10	Ph + 3-MPh → 1,8-MDPh	$2.31 \cdot 10^3$
11	Ph + 3-MPh → 3,8-MDPh	$1.91 \cdot 10^3$
12	1-MPh + 1-MPh → 1,6-MDPh + F	$1.45 \cdot 10^3$
13	1-MPh + 3-MPh → 1,8-MDPh + F	$9.17 \cdot 10^2$
14	3-MPh + 3-MPh → 3,8-MDPh + F	$5.82 \cdot 10^2$

following tables and determine the numbers used to indicate the positions occupied by the methylol groups.

The four addition reactions of formaldehyde to a diphenol are reported in Table 2.8; in this case the activation energy is assumed at the value $E_a = 90$ kJ mol^{-1} as well.

Finally, the reaction no. 89 has been included with the aim of considering, for the sake of completeness, the reactions that produce higher molecular weight molecules. In fact, as discussed before, these reactions are almost negligible in the system under study, in the sense that the system behavior is almost insensitive to such reactions at the reaction times of interest. Thus, the approximate kinetic law

$$\sum_{i=1}^{4} \mathrm{MPh}_i + \sum_{i=0}^{4} \mathrm{DPh}_i \rightarrow \text{trimers}, \tag{2.26}$$

has been assumed, with a second-order kinetics, $k_0 = 9.69 \cdot 10^3$ m^3 mol^{-1} s^{-1}, which corresponds to the average value of the other condensation reactions considered, and $E_a = 90$ kJ mol^{-1}.

Table 2.4 Condensation reactions producing the dimers DPh_1

No.	Reaction	k_0 [$m^3\,mol^{-1}\,s^{-1}$]
15	Ph + 1,5-MPh → 2-M-1,6-MDPh	$2.75 \cdot 10^4$
16	Ph + 1,5-MPh → 2-M-1,8-MDPh	$2.28 \cdot 10^4$
17	Ph + 1,3-MPh → 4-M-1,6-MDPh	$1.59 \cdot 10^3$
18	Ph + 1,3-MPh → 4-M-1,8-MDPh	$1.32 \cdot 10^3$
19	Ph + 1,3-MPh → 7-M-1,8-MDPh	$2.04 \cdot 10^3$
20	Ph + 1,3-MPh → 2-M-3,8-MDPh	$1.68 \cdot 10^3$
21	1-MPh + 1-MPh → 2-M-1,6-MDPh	$2.89 \cdot 10^3$
22	1-MPh + 1-MPh → 7-M-1,8-MDPh	$1.83 \cdot 10^4$
23	1-MPh + 3-MPh → 2-M-1,8-MDPh	$3.62 \cdot 10^2$
24	1-MPh + 3-MPh → 2-M-3,8-MDPh	$5.81 \cdot 10^3$
25	1-MPh + 3-MPh → 4-M-1,6-MDPh	$1.23 \cdot 10^4$
26	3-MPh + 3-MPh → 4-M-1,8-MDPh	$1.56 \cdot 10^4$
27	1-MPh + 1,5-MPh → 2-M-1,6-MDPh + F	$1.09 \cdot 10^4$
28	1-MPh + 1,3-MPh → 4-M-1,6-MDPh + F	$6.32 \cdot 10^2$
29	1-MPh + 1,3-MPh → 7-M-1,8-MDPh + F	$8.08 \cdot 10^2$
30	3-MPh + 1,5-MPh → 2-M-1,8-MDPh + F	$6.93 \cdot 10^4$
31	3-MPh + 1,3-MPh → 4-M-1,8-MDPh + F	$4.01 \cdot 10^2$
32	3-MPh + 1,3-MPh → 2-M-3,8-MDPh + F	$5.12 \cdot 10^2$

2.5 A General Model for a Network of Nonchain Reactions

The kinetic model developed in Sect. 2.4 for the phenol–formaldehyde reaction belongs to a wider class of kinetic networks made up of irreversible nonchain reactions. In this section, a general form of the mathematical model for this class of reactive systems is presented; moreover, it is shown that the temperature attainable in the reactor is bounded and the lower and upper bounds are computed.

To this goal, let us consider a cooled batch reactor in which the following network of irreversible reactions takes place:

$$
\begin{array}{llll}
A_1 \longrightarrow \upsilon_{1,2}A_2 & & & \\
A_1 \longrightarrow \upsilon_{1,3}A_3 & A_2 \longrightarrow \upsilon_{2,3}A_3 & & \\
\vdots & \vdots & & \\
A_1 \longrightarrow \upsilon_{1,N_C+1}A_{N_C+1} & A_2 \longrightarrow \upsilon_{2,N_C+1}A_{N_C+1} & \ldots & A_{N_C} \longrightarrow \upsilon_{N_C,N_C+1}A_{N_C+1}
\end{array}
$$

where A_i denotes the ith component, $\upsilon_{i,j} \geq 0$ is the stoichiometric coefficient of A_j in the reaction $A_i \rightarrow A_j$, and A_{N_C+1} is the final product. The above mechanism represents a general scheme of irreversible nonchain reactions involving N_C reactants and can be reduced to a simpler series and/or parallel reaction scheme by assuming $\upsilon_{i,j} = 0$ for the reactions to be eliminated.

Table 2.5 Condensation reactions producing the dimers DPh_2

No.	Reaction	k_0 [$m^3\,mol^{-1}\,s^{-1}$]
33	Ph + TMPh → 2,4-M-1,6-MDPh	$7.38 \cdot 10^3$
34	Ph + TMPh → 2,4-M-1,8-MDPh	$6.10 \cdot 10^3$
35	Ph + TMPh → 7,9-M-1,8-MDPh	$8.57 \cdot 10^3$
36	Ph + TMPh → 2,4-M-3,8-MDPh	$7.08 \cdot 10^3$
37	1-MPh + 1,5-MPh → 2,7-M-1,6-MDPh	$4.32 \cdot 10^3$
38	1-MPh + 1,5-MPh → 7,9-M-1,8-MDPh	2.39
39	1-MPh + 1,5-MPh → 2,7-M-1,8-MDPh	$6.92 \cdot 10^4$
40	1-MPh + 1,3-MPh → 2,4-M-1,6-MDPh	$7.86 \cdot 10^3$
41	1-MPh + 1,3-MPh → 2,9-M-1,6-MDPh	$2.50 \cdot 10^2$
42	1-MPh + 1,3-MPh → 4,7-M-1,8-MDPh	$4.00 \cdot 10^3$
43	1-MPh + 1,3-MPh → 2,7-M-1,8-MDPh	$3.19 \cdot 10^2$
44	1-MPh + 1,3-MPh → 2,7-M-3,8-MDPh	$5.12 \cdot 10^3$
45	3-MPh + 1,5-MPh → 2,4-M-3,8-MDPh	1.52
46	3-MPh + 1,5-MPh → 2,9-M-1,6-MDPh	$9.29 \cdot 10^4$
47	3-MPh + 1,3-MPh → 4,9-M-1,6-MDPh	$5.37 \cdot 10^3$
48	3-MPh + 1,3-MPh → 2,4-M-1,8-MDPh	$4.98 \cdot 10^3$
49	3-MPh + 1,3-MPh → 4,7-M-1,8-MDPh	$6.87 \cdot 10^3$
50	1-MPh + TMPh → 2,4-M-1,6-MDPh + F	$2.93 \cdot 10^3$
51	1-MPh + TMPh → 7,9-M-1,8-MDPh + F	$3.40 \cdot 10^3$
52	3-MPh + TMPh → 2,4-M-1,8-MDPh + F	$1.86 \cdot 10^3$
53	3-MPh + TMPh → 2,4-M-3,8-MDPh + F	$2.15 \cdot 10^3$
54	1,5-MPh + 1,5-MPh → 2,7-M-1,6-MDPh + F	$8.26 \cdot 10^4$
55	1,3-MPh + 1,5-MPh → 2,9-M-1,6-MDPh + F	$4.78 \cdot 10^3$
56	1,3-MPh + 1,5-MPh → 2,7-M-1,8-MDPh + F	$6.10 \cdot 10^3$
57	1,3-MPh + 1,3-MPh → 4,9-M-1,6-MDPh + F	$2.76 \cdot 10^2$
58	1,3-MPh + 1,3-MPh → 4,7-M-1,8-MDPh + F	$7.06 \cdot 10^2$
59	1,3-MPh + 1,3-MPh → 2,7-M-3,8-MDPh + F	$4.51 \cdot 10^2$

Assuming first-order kinetics and perfect mixing of the reactor contents, the mass balances are

$$
\begin{aligned}
\dot{C}_1 &= -k_{c1}(T_r)C_1,\\
\dot{C}_2 &= \upsilon_{1,2}k_{c1,2}(T_r)C_1 - k_{c2}(T_r)C_2,\\
\dot{C}_3 &= \upsilon_{1,3}k_{c1,3}(T_r)C_1 + \upsilon_{2,3}k_{c2,3}(T_r)C_2 - k_{c3}(T)C_3,\\
&\vdots\\
\dot{C}_{N_C} &= \upsilon_{1,N_C}k_{c1,N_C}(T_r)C_1 + \cdots\\
&\quad + \upsilon_{N_C-1,N_C}k_{cN_C-1,N_C}(T_r)C_{N_C-1} - k_{cN_C}(T_r)C_{N_C},
\end{aligned}
\tag{2.27}
$$

Table 2.6 Condensation reactions producing the dimers DPh_3

No.	Reaction	k_0 [m^3 mol^{-1} s^{-1}]
60	1-MPh + TMPh → 2,4,7-M-1,6-MDPh	$1.16 \cdot 10^3$
61	1-MPh + TMPh → 2,4,7-M-1,8-MDPh	$1.85 \cdot 10^4$
62	1-MPh + TMPh → 2,7,9-M-1,8-MDPh	$1.34 \cdot 10^3$
63	1-MPh + TMPh → 2,4,7-M-3,8-MDPh	$2.15 \cdot 10^4$
64	3-MPh + TMPh → 2,4,9-M-1,6-MDPh + F	$2.49 \cdot 10^3$
65	3-MPh + TMPh → 4,7,9-M-1,8-MDPh + F	$2.89 \cdot 10^3$
66	1,5-MPh + 1,5-MPh → 2,7,9-M-1,8-MDPh	$3.61 \cdot 10$
67	1,5-MPh + 1,3-MPh → 2,4,7-M-1,6-MDPh	$5.93 \cdot 10^4$
68	1,5-MPh + 1,3-MPh → 4,7,9-M-1,8-MDPh	1.05
69	1,5-MPh + 1,3-MPh → 2,4,7-M-3,8-MDPh	1.34
70	1,3-MPh + 1,3-MPh → 2,4,9-M-1,6-MDPh	$6.87 \cdot 10^3$
71	1,3-MPh + 1,3-MPh → 2,4,7-M-1,8-MDPh	$8.78 \cdot 10^3$
72	1,5-MPh + TMPh → 2,4,7-M-1,6-MDPh + F	$2.21 \cdot 10^4$
73	1,5-MPh + TMPh → 2,7,9-M-1,8-MDPh + F	$2.68 \cdot 10^3$
74	1,3-MPh + TMPh → 2,4,9-M-1,6-MDPh + F	$1.28 \cdot 10^3$
75	1,3-MPh + TMPh → 2,4,7-M-1,8-MDPh + F	$1.64 \cdot 10^3$
76	1,3-MPh + TMPh → 4,7,9-M-1,8-MDPh + F	$1.49 \cdot 10^3$
77	1,3-MPh + TMPh → 2,4,7-M-3,8-MDPh + F	$1.90 \cdot 10^3$

Table 2.7 Condensation reactions producing the dimers DPh_4

No.	Reaction	k_0 [m^3 mol^{-1} s^{-1}]
78	1,5-MPh + TMPh → 2,4,7,9-M-1,8-MDPh	4.48
79	1,5-MPh + TMPh → 2,4,7,9-M-3,8-MDPh	5.62
80	1,3-MPh + TMPh → 2,4,7,9-M-1,6-MDPh	$1.59 \cdot 10^4$
81	1,3-MPh + TMPh → 2,4,7,9-M-1,8-MDPh	$1.85 \cdot 10^4$
82	TMPh + TMPh → 2,4,7,9-M-1,6-MDPh + F	$5.93 \cdot 10^3$
83	TMPh + TMPh → 2,4,7,9-M-1,8-MDPh + F	$1.38 \cdot 10^4$
84	TMPh + TMPh → 2,4,7,9-M-3,8-MDPh + F	$7.98 \cdot 10^3$

where C_i $(i = 1, \ldots, N_C)$ is the concentration of the chemical species A_i, T_r is the reactor temperature, $k_{ci,j}$ $(j = 2, \ldots, N_C)$ is the rate constant of the reaction $A_i \to A_j$, following the Arrhenius law

$$k_{ci,j}(T_r) = k_{0i,j} \exp\left(-\frac{E_{ai,j}}{\mathcal{R}T_r}\right), \tag{2.28}$$

$E_{ai,j}$ is the activation energy of each reaction, and $k_{0i,j}$ is the corresponding preexponential factor. Moreover, the lumped rate constant k_{ci} of the reactions of disap-

Table 2.8 Addition reactions of formaldehyde to a dimer

No.	Reaction	k_0 [m^3 mol^{-1} s^{-1}]
85	$DPh_0 + F \rightarrow DPh_1$	$1.40 \cdot 10^5$
86	$DPh_1 + F \rightarrow DPh_2$	$2.41 \cdot 10^5$
87	$DPh_2 + F \rightarrow DPh_3$	$2.20 \cdot 10^5$
88	$DPh_3 + F \rightarrow DPh_4$	$2.52 \cdot 10^5$

pearance of species i is defined as

$$k_{ci}(T_r) = \sum_{j=i+1}^{N_C+1} k_{ci,j}(T_r). \tag{2.29}$$

In order to write the heat balance, it is assumed that all reactions are exothermic, that the temperature control is achieved by means of a heat exchange jacket surrounding the reactor, and that heat losses to the environment can be neglected. Energy balances can be written both for the fluid in the reactor and for the fluid in the jacket. In the first case, by including in (2.25) the proper expression for the total heat of reaction, this equation reads

$$\dot{T}_r = \frac{\sum_{i=1}^{N_C} \sum_{j=i+1}^{N_C+1} (-\Delta H_{Ri,j}) k_{ci,j}(T_r) C_i}{\rho_r c_r} - \frac{US(T_r - T_j)}{\rho_r c_r V_r}, \tag{2.30}$$

where $\Delta H_{Ri,j}$ denotes molar enthalpy change of each reaction.

In the second case, for the sake of simplicity, the jacket is modeled as a continuous well-mixed vessel, so that the general framework of the mass balance for a CSTR (2.23) can be adequately rearranged to give

$$\dot{T}_j = \frac{F_V(T_{in} - T_j)}{V_j} + \frac{US(T_r - T_j)}{\rho_j c_j V_j}, \tag{2.31}$$

where the subscript j denotes variables referred to the jacket, F_V and T_{in} are the volume flow rate and the temperature of the fluid entering the jacket, respectively.

All the stages of the reaction cycle (i.e., initial reactor heating, reaction development, and final quenching) can be described by (2.30) and (2.31). Indeed, the second term on the right-hand side of (2.30) usually turns out to be negative during the heating phase (when $T_{in} > T_j > T_r$) and positive during temperature control and final cooling ($T_{in} < T_j < T_r$).

It can be easily recognized that the rate constants are nonnegative and strictly increasing functions of the reactor temperature T_r. Since the reaction is assumed to be exothermic and T_{in} is bounded, i.e., $T_{in,min} \leq T_{in} \leq T_{in,max}$, the temperature in the reactor is lower bounded by the value

$$T_{r,min} = \min\{T_{r,0}, T_{j,min}\},$$

where $T_{\mathrm{r},0}$ is the initial reactor temperature, and $T_{\mathrm{j,min}}$ is the minimum attainable jacket temperature, which coincides with the minimum attainable value, $T_{\mathrm{in,min}}$, of T_{in}. Moreover, extending the concept of adiabatic reaction temperature introduced in Sect. 2.3.2, an upper bound for T_r can be computed by considering the ideal heating/reaction scheme composed by the following two steps:

- the reacting mixture is first heated up to the maximum temperature value, $T_{\mathrm{in,max}}$, of the fluid entering the jacket; and
- then the complete reactants conversion takes place adiabatically.

The numerical value of the upper bound is given by

$$T_{\mathrm{r,max}} = T_{\mathrm{in,max}} + C_{10}\frac{(-\Delta H_{\mathrm{R}1,N_\mathrm{C}+1})}{\rho_\mathrm{r} c_{p\mathrm{r}}},$$

where C_{10} is the initial concentration of A_1. As a consequence, the rate constants are also bounded as follows:

$$0 < \underline{k}_{\mathrm{c}i,j} \le k_{\mathrm{c}i,j}(T_\mathrm{r}) \le \overline{k}_{\mathrm{c}i,j} \quad \forall\, T_\mathrm{r},\ i = 1,\dots,N_\mathrm{C},\ j = i+1,\dots,N_\mathrm{C}+1, \tag{2.32}$$

where $\underline{k}_{\mathrm{c}i,j} = k_{\mathrm{c}i,j}(T_{\mathrm{r,min}})$ and $\overline{k}_{\mathrm{c}i,j} = k_{\mathrm{c}i,j}(T_{\mathrm{r,max}})$.

In view of inequalities (2.32), also the lumped rate constants $k_{\mathrm{c}i}$ can be bounded as follows:

$$0 < \underline{k}_{\mathrm{c}i} \le k_{\mathrm{c}i}(T_\mathrm{r}) \le \overline{k}_{\mathrm{c}i} \quad \forall T_\mathrm{r},\ i = 1,\dots,N_\mathrm{C}, \tag{2.33}$$

where $\underline{k}_{\mathrm{c}i} = \sum_{j=i+1}^{N_\mathrm{C}+1} \underline{k}_{\mathrm{c}i,j}$ and $\overline{k}_{\mathrm{c}i} = \sum_{j=i+1}^{N_\mathrm{C}+1} \overline{k}_{\mathrm{c}i,j}$.

These results will be used in Chap. 5 when dealing with model-based control of nonchain reactions in a cooled batch reactor.

2.6 Measuring the Reactor Status

The modeling approach to the batch reactor presented in the previous sections of this chapter must be strengthened and extended by considering the relationship between the user and the reactor. This relationship may be divided into two different parts, namely the methods for measuring the reactor status and the actions to be taken in order to change it.

Measuring the variables which define the reactor status is important both in the laboratory and in the industrial practice. In the first case, measuring gives the experimental information necessary to tune the mathematical models, i.e., to determine the values of the adjustable parameters, a task to whom the entire Chap. 3 is devoted. In industry, measuring and regulating the operative conditions is very important in order to ensure an adequate quality of the final product and safe operation of the batch cycle. These two ways of using measurements pose different problems to the user, because they require different properties of the measurement device. Thus, before considering in some detail the main measurable variables, a short review of the measuring qualities is deemed to be useful.

2.6.1 Measurements Quality

In a few simpler cases, the use of very basic measuring instruments involves the direct comparison of the physical quantity to be measured with predisposed samples and scales, as happens, for example, when measuring a length with a ruler. On the contrary, almost all the measurements of industrial interest are obtained by means of a more complex measuring instrument, which is a device that translates some physical effect depending on the measurable quantity in a signal usable at the user interface.

In general, the relationship between input and output of a measuring instrument is a complex function that must be determined by calibration, i.e., by recording the output corresponding to different known values of the input, so that a *calibration curve* can be obtained. In most cases, the functionality represented by the calibration curve is a mere interpolation of experimental points rather than a predetermined theoretical law. The slope of the calibration curve defines the *sensitivity* of the measuring instrument; in general, the higher the sensitivity, the more accurate is the measurement.

The quality of measurements is determined by a compromise between costs and process reliability and can be described in terms of accuracy, precision or repeatability, resolution, response time, and stability. The first two properties are strictly related to the concept of measurement error, which can never be completely eliminated. The *accuracy* of a measuring instrument is the ability to provide outputs which are, on average, close to the true value of the underlying variable. The *precision* (or repeatability) of a measuring instrument is the ability to replicate the measured values under the same conditions and corresponds to the standard deviation of the errors. The *resolution* is the ability of a measuring system to detect small changes in the measured variable; the smaller the changes detected by the instrument, the higher its resolution. The *response time* of an instrument can be defined as the time needed to let the instrument output settle around the measured value for an input step change. Finally, *stability* (or ruggedness) is the ability of a sensor to maintain its performance under the expected operative conditions.

2.6.2 Online Measurements

In order to ensure an adequate quality of products and a safe operation, the monitoring of a batch reactor should include, at least, online measurements of temperature, pressure, and of some composition-related variables. In this context, online measurements may be defined as measurements obtained via instruments strictly connected to the reactor and characterize by response times markedly smaller than the characteristic times of the chemical reaction. In general, this is the case of temperature and pressure, which can be easily measured online by means of reliable, relatively cheap, and poorly intrusive sensors. This allows the introduction of sensor redundancy, a common practice to increase reliability. On the other hand, online

Table 2.9 Comparison of RTD and thermocouple characteristics

	RTD	Thermocouple
Temperature Range [°C]	−200 to 850	−200 to 1800
Accuracy [°C]	10^{-4}–10^{-2}	0.1–2
Response Time [s]	0.5–5	0.01–1
Stability	High	Moderate
Sensitivity	High	Low
Cost	Moderate	Low

measurements of chemical composition is a more challenging task, especially in liquid systems, so that indirect measurements through composition related overall properties (e.g., pH) often represent the only viable alternative.

Temperature is by far the more frequently measured state variable and is considered in some detail hereafter. Basic temperature measurement in a batch reactor must regard, at least, the reacting mixture and the heat exchange fluid. To this goal, the devices most widely used are thermocouples and resistance temperature detectors (RTD).

Thermocouples are based on the thermoelectric Seebeck effect, which generates a voltage at the junction between two metallic conductors, which depends on temperature [13]. Thus, in the measuring circuit, two junctions are created, namely, a sensitive (or hot) junction at the point where temperature has to be measured and a nonsensitive (cold) junction, kept at a constant known temperature, where the voltage established between the conductors can be easily measured [19]. Different typologies of thermocouples exist for application in a wide range of conditions; they essentially differ by the materials, the most common being J (iron/constantan), K (chromel/alumel), T (copper/constantan), and E (chromel/constantan).

RTD sensors consist of a platinum (with copper and nickel being cheaper alternatives) wire wrapped in a coil and traversed by a constant current, typically in the range 0.8–1.0 mA. Since the electric resistance of these materials changes almost linearly with temperature at a rate of about 0.3%°C^{-1}, the voltage drop across the sensor is easily converted into a temperature value. Accuracy of RTDs is on average higher than that of thermocouples, and platinum-based resistive detectors can be as accurate as 10^{-4}°C [19]. On the other hand, their response time is somewhat larger, because of the time needed by the measuring coil to reach thermal equilibrium with the surroundings. A good alternative to RTDs are thermistors, based on the resistance sensitivity of semiconductors with respect to temperature.

Comparison of RTD and thermocouple average performance is reported in Table 2.9. Here, the response time is defined as the time needed by the device to reach the final temperature within 0.5% of its value for a temperature step change.

Besides the electrical devices described above, mechanical systems for temperature measurement are not uncommon in chemical reactors. As an example, systems consisting in a bulb connected to a temperature-sensitive volume or pressure element can occasionally offer some interesting advantages, e.g., independence from external electrical power and simpler maintenance [2].

The optimal location of temperature sensors is not a trivial task, because, in industrial scale chemical reactors, temperature gradients can be generated by non-perfect mixing and by the heat exchange device. In systems that are very sensitive to temperature (as in the case of the phenol–formaldehyde reaction introduced in Sect. 2.4), the temperature should be measured in the hottest point (*hot spot*), a strategy which proves to be too conservative when safety is not an issue, since the reaction rate in the remaining parts of the reactor could prove to be rather slow. Moreover, it should be considered that the hot spots may move within the reactor according to the different stages of operation.

A similar problem affects the heat exchange jacket and may be reduced by using a coil, in which a plug flow can be assumed for the heat exchange fluid. When the reaction temperature is controlled by an external heat exchanger or condenser [17], the recirculation of the fluids introduces a transport delay that may strongly affect the control action.

It is also interesting to briefly consider online measurements of variables different from temperature [5]. Since pressure is defined as the normal force per unit area exerted by a fluid on a surface, the relevant measurements are usually based on the effects deriving from deformation of a proper device. The most common pressure sensors are piezoresistive sensors or strain gages, which exploit the change in electric resistance of a stressed material, and the capacitive sensors, which exploit the deformation of an element of a capacitor. Both these sensors can guarantee an accuracy better than 0.1 percent of the full scale, even if strain gages are temperature sensitive.

Pressure is more directly connected to the concept of explosion; nevertheless, it is less directly connected to the reactor status, since, for liquid-phase reactors, pressure nonlinearly depends on temperature (trough the vapor pressure relationship) and concentration (through the activity coefficients in liquid phase). Moreover, since pressure measurements are usually less accurate than temperature measurements, they are to be considered in particular for gassy reactions, i.e., when the runaway produces small temperature effects but large amounts of incondensable products in gaseous phase.

Online measurements of composition would be very appreciated, because composition is the most important variable. Unfortunately, direct measurements of the amount of a single component can be obtained only in a few cases, and typically for gaseous system, an example being the measurement of oxygen based on its paramagnetism. In fact, liquid phase systems are usually made out of components of similar chemical structure, and these must be separated before measuring their quantity.

Thus, online measurements of composition are usually limited to some overall property. A typical example is pH, defined as the absolute value of the logarithm of the molar concentration (or, more exactly, activity) of hydrogen ion; pH can be measured by exploiting the electric potential established between two proper electrodes immersed in the sample fluid, usually a glass membrane electrode and a reference electrode [15]. Notwithstanding the temperature dependence and the alkaline error (at high pH, a marked sensitivity to the effect of Na^+ and of other monovalent cations), accuracies of 0.01–0.1 pH units can be achieved. Alternatively, electric conductivity and optical density may give some online information about composition [1].

Finally, one can easily obtain the properties of the stirring system (rotational speed and torque) and compute the stirring power from their product, but these variables are only indirectly linked to the reactor status through the system density.

2.6.3 Offline Measurements

An offline measurement apparatus is usually not directly mounted on the reactor, but is fed with samples withdrawn from it manually or automatically. This is the typical case of chromatography, a widely used measurement device for gas and liquid composition. Both gas and liquid chromatographies are based on the separation of the sample by means of selective adsorption on a solid substrate posed in a fixed bed column, and on the detection of the change of a suitable property of the (gas or liquid) carrier, usually thermal conductivity.

It appears that each component is characterized by a retention time (which also depends on the substrate and on the column length and temperature) and by a relationship between its amount and the thermal conductivity of the modified carrier. Therefore, not only a calibration curve is required for any component, but also the operating conditions must be optimized in order to obtain a sharp separation among the different components.

In conclusion, a complete analysis of a complex mixture can require very long times (from a few minutes up to many hours); thus, such a measuring apparatus is not suited for online measurements to be used in reactor control. On the contrary, in the laboratory, chromatography is very often the preferred method of analysis of complex mixtures, since these more accurate data can be used to identify the reaction mechanism and the relevant kinetic parameters.

This brief overview of offline measurements can be concluded by considering the measurements of the heat released by chemical reactions, which can be obtained via calorimetric measurements [7, 18]. The most diffused industrial calorimeters are the so-called reaction calorimeters, basically consisting in jacketed vessels in which the reaction takes place and the heat released is measured by monitoring the temperature of the fluid in the jacket. A class of alternative instruments are the scanning calorimeters (differential or adiabatic), in which the analysis is performed by linearly increasing the sample temperature with respect to time, in order to test the reactivity of potentially unstable chemical systems in a proper temperature range by measuring the released heat.

2.7 Manipulating the Reactor Status

Chemical reactions proceed at very different reaction rates, so that typical values of characteristic reaction time in industrial batch reactors range from few hours to several days. The largest values are encountered in chemical systems like polymerization reactions, the smallest in some biological reactions such as fermentation of sugars to alcohols. Thus, a challenging problem of automatic control is met in the first case, whereas, in the second case, the reaction is often controlled manually, and the problem is restricted to selecting the best action to be executed in a given state of the process.

In both cases, it is of interest to single out the variables that can be manipulated; it is possible to distinguish the actions scheduled in the design of the process, the actions aimed at controlling the reactor in the presence of failures, and the emergency actions to be operated in the case of runaway.

In general, it does not make much sense to modify the speed of the mechanical stirrer, since its design value is fixed to optimize the mixing of the reactor contents and the value of the heat transfer coefficient. On the contrary, it is possible to modify the reaction rate and consequently the temperature by adding to the reacting mixture a proper amount of fresh solvent, and/or of a reactant, a catalyst, or a reaction inhibitor.

In many cases, these operations are scheduled as standard operating procedures; nevertheless, the addition of fresh solvent and/or of a reaction inhibitor can be also used as an emergency protection measure against explosions. This suppression system should be preferred to the alternative method consisting in a bursting disc, which must be provided with a discharge line to an emergency tank, since the discharge into the environment of the reactor contents must be avoided. In fact, the design of the discharge line in the presence of unsteady two-phase flashing flow is not straightforward, whereas the only drawback of a suppressing system is the need for allowing a larger gaseous head in the reactor.

In this book, automatic control of reactor temperature is the most interesting target; to this purpose, the manipulated variable is usually one (or a combination of) the following [3]:

- the flow rate of the heat exchange fluid
- the inlet temperature of the heat exchange fluid; and
- the heat exchange surface.

The first strategy appears to be very effective; in fact, as shown by (2.30) and (2.31), the direct (linear) effect of increasing the flow rate is augmented by the increase of the jacket-side heat exchange coefficient. This control action can be realized without a noticeable time delay by a simple control valve, the only drawback for its quantitative assessment being the nonlinear relationship between the overall heat exchange coefficient and the flow rate of the heat exchange fluid.

The inlet temperature of the heat exchange fluid has a more easily predictable effect on the system dynamics (in fact, the small change of the heat transfer coefficient with temperature is usually negligible) and is therefore preferred in model-based control approaches. The implementation of this strategy requires the availability of two compatible heat exchange fluids (usually water for cooling and steam for heating) and of a mixing device for conditioning the heat exchange fluid at the required flow rate and temperature, which introduces a time delay in the system.

Finally, the choice of the heat exchange surface area as adjustable parameter is a recent concept developed after the diffusion of cooling systems partitioned into separated compartments. By controlling the corresponding opening/closing valves, heat exchange compartments can be excluded or joined, whereas inlet temperature and flow rates are kept constant. In this way, the exchange surface can be adjusted according to the degree of filling of the reactor and to the heat exchange flux required for temperature control. Other advantages are the fast response of the cooling system to the control action and the absence of a pronounced cold/hot region at the reactor wall since the coolant inlets are manifold. Drawbacks of this configuration are the higher pressure drops and the more complex flow patterns of the heat exchange fluid, which affect the modeling approach.

2.8 Conclusions

The modeling of chemical batch reactors has been chosen as the starting point for the roadmap developed in this book. The simplified mathematical models presented in the first sections of the chapter allow us to focus the attention on different aspects of chemical kinetics, whereas the causes of nonideal behavior of chemical batch reactors are faced in the last chapter.

The rather complex issue of chemical kinetics has been discussed in a quantitative way, in order to stress out two main ideas, namely, the necessity of resorting to simplified kinetic models and the need of adequate methods of data analysis to estimate the kinetic parameters. These results introduce Chap. 3, in which basic concepts and up-to-date methods of identification of kinetic parameters are presented.

To this purpose, a brief overview of the measurable variables in a batch reactor has been included; the difference between online measurements, suitable for control purposes, and offline measurements, which can be exploited to obtain experimental data to be used for the identification, has been stressed.

In the second part of the chapter, the mathematical model of the BR has been augmented by considering its behavior in the presence of significant thermal effects and of a proper heat exchange apparatus. In particular, modeling these aspects brings the reader to understand the need for considering the thermal stability of batch reactors (Chap. 4) and the need for adequate systems of automatic temperature control (Chap. 5).

References

1. J.H. Altman. Densitometry measurements. In J.C. Webster, editor, *Measurement, Instrumentation, and Sensors Handbook, 2nd Edition*, Chap. 57. CRC Press, Boca Raton, 1999.
2. N.A. Anderson. *Instrumentation for Process Measurement and Control*. CRC Press, Boca Raton, 1998.
3. D. Bonvin. Optimal operation of batch reactors—a personal view. *Journal of Process Control*, 8(5/6):355–368, 1998.

4. M. Buback and A.M. van Herk. *Radical Polymerization: Kinetics and Mechanism*. Wiley-VCH, Weinheim, 2007.
5. K.H.L. Chau, R. Goehner, E. Drubetsky, H.M. Brady, W.H. Bayles, Jr., and P.C. Pedersen. Pressure and sound measurement. In J.C. Webster, editor, *Measurement, Instrumentation, and Sensors Handbook, 2nd Edition*, Chap. 26. CRC Press, Boca Raton, 1999.
6. I. Chorkendorff and J.W. Niemantsverdriet. *Concept of Modern Catalysis and Kinetics*. Wiley-VCH, Weinheim, 2003.
7. J.A. Dean. *The Analytical Chemistry Handbook*. McGraw Hill, New York, 1995.
8. K.C. Eapen and L.M. Yeddenapalli. Kinetics and mechanism of the alkali-catalyzed addition of formaldehyde to phenol and substituted phenols. *Die Makromolekulare Chemie*, 119(2766):4–16, 1968.
9. W. Hesse. Phenolic resins. In *Ullmann's Encyclopedia of Industrial Chemistry, 6th Edition*. Wiley, New York, 2001.
10. O. Levenspiel. *Chemical Reaction Engineering, 3rd Edition*. Wiley, New York, 1999.
11. L.B. Manfredi, C.C. Riccardi, O. de la Osa, and A. Vazquez. Modelling of resol resin polymerization with various formaldehyde/phenol molar ratios. *Polymer International*, 50:796–802, 2001.
12. J. Nielsen, J. Villadsen, and G. Lidén. *Bioreaction Engineering Principles, 2nd Edition*. Kluwer Academic, New York, 2003.
13. R.P. Reed. Thermal effects in industrial electronics circuits. In J.D. Irwin, editor, *Industrial Electronics Handbook*. CRC Press, Boca Raton, 1996.
14. C.C. Riccardi, G. Astarloa Aierbe, J.M. Echeverria, and I. Mondragon. Modelling of phenolic resin polymerisation. *Polymer*, 43:1631–1639, 2002.
15. N.F. Sheppard, Jr. and A. Giuseppi-Elie. PH measurements. In J.C. Webster, editor, *Measurement, Instrumentation, and Sensors Handbook, 2nd Edition*. Chap. 71. CRC Press, Boca Raton, 1999.
16. G.P. Smith, D.M. Golden, M. Frenklach, N.W. Moriarty, B. Eiteneer, M. Goldenberg, C.T. Bowman, R.K. Hanson, S. Song, W.C. Gardiner Jr., V.V. Lissianski, and Z. Qin. GRI-Mech 3.0. A kinetic mechanism to model natural gas combustion. www.me.berkley.edu/grimech/.
17. P. Trambouze and J.-P. Euzen. *Chemical Reactors: From Design to Operation*. Editions Technip, Paris, 2004.
18. S. van Herwaarden. Calorimetry measurement. In J.C. Webster, editor, *Measurement, Instrumentation, and Sensors Handbook, 2nd Edition*, Chap. 36. CRC Press, Boca Raton, 1999.
19. J.G. Webster, editor. *Measurement, Instrumentation, and Sensors Handbook*. CRC Press, Boca Raton, 1999.

Chapter 3
Identification of Kinetic Parameters

List of Principal Symbols

A	reactant phenol
C	concentration [$\mathrm{mol\,m^{-3}}$]
Co	confirmation of a theory
d	experimental data
$\boldsymbol{D}$	matrix of experimental data
e	experimental error
E_a	activation energy [$\mathrm{J\,mol^{-1}}$]
Ex	experimental results
$\boldsymbol{f}$	function in an implicit mathematical model
f_p	probability density function
$\boldsymbol{G}$	positive definite matrix
$\boldsymbol{H}$	Hessian matrix
ΔH_R	molar enthalpy change of reaction [$\mathrm{J\,mol^{-1}}$]
I	reaction intermediate
$\boldsymbol{I}$	identity matrix
k_0	preexponential factor [$(\mathrm{mol\,m^{-3}})^{1-n}\,\mathrm{s^{-1}}$]
k_c	rate constant [$(\mathrm{mol\,m^{-3}})^{1-n}\,\mathrm{s^{-1}}$]
n	reaction order
N_C	number of dependent or state variables or of components
N_D	number of data in the sample
N_L	number of lumped chemical reactions
N_M	number of measured variables
N_P	number of adjustable parameters
N_U	number of input variables or constants
N_Z	number of isothermal runs
p	probability
P	desired product, trimethylolphenol
$\dot{q}$	specific thermal power [$\mathrm{J\,m^{-3}\,s^{-1}}$]
R	reaction rate [$\mathrm{mol\,m^{-3}\,s^{-1}}$]
$\mathcal{R}$	universal gas constant [$\mathrm{J\,mol^{-1}\,K^{-1}}$]

F. Caccavale et al., *Control and Monitoring of Chemical Batch Reactors*,
Advances in Industrial Control,
DOI 10.1007/978-0-85729-195-0_3,

$\hat{s}_{\mathrm{D}}^2$ corrected sample variance
T temperature [K]
Th theory
U objective function
$\boldsymbol{u}$ vector of input variables
$\boldsymbol{V}$ covariance
$\boldsymbol{x}$ vector of state variables
$\boldsymbol{y}$ vector of measured state variables (outputs)
$\boldsymbol{Y}$ matrix of computed values to be compared with the experimental data
$y_{\dot{q}}$ computed value of the specific thermal power [$\mathrm{J\,m^{-3}\,s^{-1}}$]
w weights
W undesired product

Greek Symbols

α constant in (3.66)
γ coefficient in (3.31)
$\boldsymbol{\Gamma}$ matrix of coefficients in (3.32)
ϵ error generated by the model
ζ generic random variable
θ adjustable parameter
$\boldsymbol{\theta}$ vector of adjustable parameters
κ step length
λ damping factor
ν corrective factor for the Levenberg–Marquardt method
σ^2 universe variance
σ_C root mean squared errors for the concentrations
$\sigma_{\dot{q}}$ root mean squared errors for the specific thermal power
φ partial sensitivity
$\boldsymbol{\phi}$ function in an explicit mathematical model
$\boldsymbol{\Phi}$ function in a linear model
ψ known term in (3.31)
$\boldsymbol{\Psi}$ matrix of known terms in (3.33)

Subscripts and Superscripts

av mean value
m measured value
max maximum
min minimum
r reactor
s step index in the nonlinear optimization procedure
$\hat{}$ best estimate or optimal value
o reference value

3.1 Bayesian Approach and Popper's Falsificationism

In science and engineering an improvement of knowledge is derived from the comparison between theories and experiments. While theories give the general framework for human understanding, experiments represent absolutely necessary tests for establishing the required correspondence between theories and real world. In most cases, the properties of the world are tentatively described by one or more alternative theories that generally contain one or more parameters not exactly known, so that they may be considered to be adjustable. Thus, the expected results are the identification of the best available theory and the evaluation of the best estimates of the relevant adjustable parameters.

When the significance and the reliability of the correspondence between theories and experiments are considered, two main alternatives are available. The *Standard View*, based on the ideas of logical empiricism, assumes that the experiments can confirm a scientific theory, i.e., that they can increase its probability (here intended as logical confidence in its truth, i.e., in its correspondence with the real world). On the contrary, *Falsificationism*, first proposed by Karl Popper [17], claims that experiments cannot demonstrate the truth of a theory but can only falsify the theory, i.e., demonstrate that a theory is unfit to describe an experimental result.

In its simpler form, the idea of falsification can be derived from the laws of Logic; in fact, if Th stands for *theory* and Ex for *experimental result*, the *modus tollens* gives

$$\big((\mathrm{Th} \rightarrow \mathrm{Ex}) \cap \neg\mathrm{Ex}\big) \rightarrow \neg\mathrm{Th}, \tag{3.1}$$

which reads: if an experimental result foreseen by a theory is not observed, the theory can be considered to be false. Moreover, since logical implication is not symmetric with respect to negation, the observation of the experimental result does not determine the truth of the theory; in *formulae*,

$$\neg\big(((\mathrm{Th} \rightarrow \mathrm{Ex}) \cap \mathrm{Ex}) \rightarrow \mathrm{Th}\big). \tag{3.2}$$

When only two alternatives are possible (Th or ¬Th), it is possible to demonstrate a theory Th from the falsity of a consequence of ¬Th, as in the *ex absurdo* demonstrations used in mathematics. Unfortunately, things are not so simple in the physical world, where *ex absurdo* demonstrations are not acceptable since the alternatives to Th may be infinite (and actually not all known); therefore, we can never rely on a theory to be a definitive description of reality.

Moreover, in the physical world, the observed experimental results are biased by the presence of experimental errors, so that the concept of truth must be intended in a less strong sense. Finally, a theory Th is not a simple statement but rather a complex set of hypotheses (in some case not all explicitly stated, but situated in the background of our knowledge). If the theory disagrees with experiments, which of the hypotheses is falsified by the experiment?

The above considerations show that the absolute evaluation of a theory is a very ambiguous task; however, in most cases, experimental data analysis is performed in order to compare two or more alternative theories, so that only a relative evaluation of quality is required. In fact, when the performance of two alternative theories is compared, it is possible to obtain a classification of quality at the actual level of experimental information, so that one of the theories can be preferred, and another discarded. In this relative sense, it is not very meaningful if the chosen theory is confirmed by experiments or if the rejected theory is falsified. Also considering that Popper was never able to present a quantitative version of falsificationism, we can proceed following the standard view, i.e., the method of comparison based on the Bayesian approach.

The concept of conditioned probability was originally developed by Thomas Bayes [3] in terms of frequencies of occurrence of events, but it can be adapted to the logical probability, i.e., to probability intended as logical confidence. Thus, in the well-know definition

$$p(\mathrm{Th} \mid \mathrm{Ex}) = \frac{p(\mathrm{Th} \cap \mathrm{Ex})}{p(\mathrm{Ex})}, \tag{3.3}$$

the a posteriori probability $p(\mathrm{Th} \mid \mathrm{Ex})$ represents the confidence in the theory Th on the basis of the observed experimental result Ex. Equation (3.3) can be written in a more useful form by exchanging Th and Ex,

$$p(\mathrm{Ex} \mid \mathrm{Th}) = \frac{p(\mathrm{Ex} \cap \mathrm{Th})}{p(\mathrm{Th})}, \tag{3.4}$$

and by recognizing that $p(\mathrm{Th} \cap \mathrm{Ex}) = p(\mathrm{Ex} \cap \mathrm{Th})$. In this way, the Bayes theorem can be obtained in the form

$$\frac{p(\mathrm{Th} \mid \mathrm{Ex})}{p(\mathrm{Th})} = \mathrm{Co}(\mathrm{Th}, \mathrm{Ex}) = \frac{p(\mathrm{Ex} \mid \mathrm{Th})}{p(\mathrm{Ex})}. \tag{3.5}$$

The left hand side of (3.5) is the ratio between the confidence in the theory obtained a posteriori, i.e., after the analysis of the experimental data, and the relevant confidence a priori; this ratio may be assumed to be a measure of the confirmation Co(Th, Ex) that the theory receives from the data. The right-hand side of (3.5) is the ratio between the a posteriori probability of the experimental results, $p(\mathrm{Ex} \mid \mathrm{Th})$, the so-called likelihood of the data, and the relevant a priori probability, $p(\mathrm{Ex})$.

While it is difficult to give a quantitative absolute value to $p(\mathrm{Ex})$, and thus to Co(Th, Ex), it appears that, if the comparison between two theories is pursued, the knowledge of $p(\mathrm{Ex})$ is no more necessary, since the ratio of confirmations is expressed as a simpler ratio of likelihoods,

$$\frac{\mathrm{Co}(\mathrm{Th}_1, \mathrm{Ex})}{\mathrm{Co}(\mathrm{Th}_2, \mathrm{Ex})} = \frac{p(\mathrm{Ex} \mid \mathrm{Th}_1)}{p(\mathrm{Ex} \mid \mathrm{Th}_2)}. \tag{3.6}$$

In order to apply (3.6) to real problems such as the identification of a kinetic model, not only the concept of likelihood must be defined more quantitatively, but also the concepts of *experiments* and of *theory* must be adapted as discussed in the next section.

3.2 Experimental Data and Mathematical Models

All experimental measurements are affected by errors. In general, experimental errors are made out of systematic errors and random errors. Systematic errors show a dependence on the operating conditions and may be caused, e.g., by calibration errors of sensors. Since these errors are absent in a well-performed experimental campaign and can be corrected by an improved experimental practice, they are not considered any more in this context.

On the other hand, random errors do not show any regular dependence on experimental conditions, since they are generated by many small and uncontrolled causes acting at the same time, and can be reduced but not completely eliminated. Thus, random errors are observed when the same measurement is repeatedly performed. In the simplest case, the *universe* of random errors is described by a continuous random variable e following a normal distribution with zero mean, i.e., for a univariate variable, the probability density function is given by

$$f_P(e) = \frac{\exp(-\frac{0.5e^2}{\sigma^2})}{\sqrt{2\pi}\sigma}. \tag{3.7}$$

This function contains the second moment or variance, σ^2, which measures the dispersion of the data around the mean. Repeated measurements give a sample of N_D elements of this universe; the methods of inverse inference allow one to evaluate the properties of the universe from the properties of the sample [10].

First, the experimental errors e_j ($j = 1, \dots, N_\mathrm{D}$) can be evaluated assuming the data average d_av as the true value (i.e., by setting $e_j = d_j - d_\mathrm{av}$, so that the mean of errors, $e_\mathrm{av} = 0$, is assumed to be equal to the expected universe mean); then, the corrected sample variance $\hat{s}_\mathrm{D}^2$ can be used as an estimate of the universe variance σ^2:

$$\sigma^2 \simeq \hat{s}_\mathrm{D}^2 = \frac{1}{N_\mathrm{D} - 1} \sum_{j=1}^{N_\mathrm{D}} e_j^2. \tag{3.8}$$

Here, the term $N_\mathrm{D} - 1$ represents the residual degrees of freedom of the sample after the estimation of the expected universe mean, through the computation d_av.

When N_M different variables are measured, a different variance σ_m^2 ($m = 1, \dots, N_\mathrm{M}$), is expected for each measured variable, since the experimental methods of measure and the relevant accuracies may be significantly different. In most cases it is reasonable to assume the same accuracy in each experiment, i.e., $\sigma_{m,j}^2 = \sigma_m^2 \ \forall j$; for the sake of simplicity, this simplified hypothesis is assumed in the following.

It is also necessary to establish whether the measurements are independent or not. This question arises in the case of multivariate random variables, i.e., when more than one variable is measured in each experiment, and, less frequently, for the measurements of the same variable in different experiments. For a given couple of random variables, ζ_1 and ζ_2, the independency of data is measured by the covariance, which extends the concept of variance,

$$\operatorname{cov}(\zeta_1, \zeta_2) = \frac{1}{N_D} \sum_{j=1}^{N_D} (\zeta_{1,j} - \zeta_{1,\mathrm{av}})(\zeta_{2,j} - \zeta_{2,\mathrm{av}}), \tag{3.9}$$

where the subscript av, as before, denotes the mean value.

Theories are not used directly, as in the discussion presented in Sect. 3.1, but allow building a mathematical model that describes an experiment in the unambiguous language of mathematics, in terms of variables, constants, and parameters. As an example, when considering the identification of kinetic parameters of chemical reactions from isothermal experiments performed in batch reactors, the relevant equations of mass conservation (presented in Sect. 2.3.1) give a set of ordinary differential equations in the general form

$$\dot{\boldsymbol{x}}(t) = \boldsymbol{f}\big(t, \boldsymbol{x}(t), \boldsymbol{u}(t), \boldsymbol{\theta}\big), \tag{3.10}$$

where the dot indicates the time derivative, $\boldsymbol{x}$ is the $(N_C \times 1)$ vector of state variables (which contains the concentrations of the different components included in the kinetic scheme), and $\boldsymbol{f}$ is an $(N_C \times 1)$ vector function. Moreover, in (3.10), $\boldsymbol{u}$ is an $(N_U \times 1)$ vector of inputs, determining the experimental conditions, and $\boldsymbol{\theta}$ is an $(N_P \times 1)$ vector of parameters.

The time variable t, the initial time t_0, the values of the input $\boldsymbol{u}(t)$ for $t \in [t_0, t[$, and the initial state $\boldsymbol{x}_0 = \boldsymbol{x}(t_0)$ are assumed to be known without noticeable experimental errors. On the other hand, the parameters are to be considered constants, whose value are unknown (or, at least, not exactly known) a priori. The parameters can be changed in a suitable range in order to analyze the model sensitivity, i.e., its capability to describe different scenarios, or can be adjusted, in order to improve the correspondence between the experimental data and the computed variables; in chemical kinetics, this is usually the case of rate constants and of activation energies.

Equation (3.10) can be integrated in a few simple cases, thus producing the explicit form

$$\boldsymbol{x}(t) = \boldsymbol{\phi}(t, t_0, \boldsymbol{x}_0, \boldsymbol{u}_{[t_0,t[}, \boldsymbol{\theta}), \tag{3.11}$$

where $\boldsymbol{u}_{[t_0,t[}$ denotes the set of values taken by the input in the time interval $[t_0, t[$. Equation (3.11) allows computing the vector $\boldsymbol{x}$ at any time instant from the $(N_C \times 1)$ vector function $\boldsymbol{\phi}$; in more complex cases, discussed in some detail in Sect. 3.6, the vector $\boldsymbol{x}$ must be computed via numerical integration of the implicit model (3.10). Usually, in chemical kinetics, the explicit models (3.11) are nonlinear functions of the parameters; nevertheless, linear-in-the-parameter models are briefly considered in Sect. 3.4 because of their general relevance.

In conclusion, for any assigned value of $\boldsymbol{\theta}$, mathematical models produce a set of computed results $\boldsymbol{x}(\boldsymbol{\theta})$, which must be compared to the corresponding experimental values. In general, only the concentrations of N_M components are measured at N_D experimental times t_j, yielding the $(N_M \times N_D)$ matrix $\boldsymbol{D} = \{d_{m,j}\}$ of the experimental data. Thus, a subset of $\boldsymbol{x}$, the $(N_M \times 1)$ vector of outputs $\boldsymbol{y}$, must be used in the comparison. The computed values corresponding to $\boldsymbol{D}$ are grouped in the $(N_\mathrm{M} \times N_\mathrm{D})$ matrix $\boldsymbol{Y} = \boldsymbol{Y}(\boldsymbol{\theta}) = \{y_{m,j}\}$, where $\boldsymbol{\theta}$ is the current estimate of the vector of parameters.

This comparison never shows a perfect correspondence between models and experiments because of modeling and measurement errors. In fact, even if the presence of systematic experimental errors can be excluded, systematic errors generated by the inadequacy of the model must be added to random experimental errors; for each measured variable ($m = 1, \dots, N_\mathrm{M}$) and each experimental time ($j = 1, \dots, N_\mathrm{D}$), the errors generated by the model are defined as

$$\epsilon_{m,j} = d_{m,j} - y_{m,j}. \tag{3.12}$$

On the other hand, since the model must be considered as a flexible tool that can be adapted to the experimental data by changing the values of the adjustable parameters, the method consists in computing the optimal values of the parameters, $\widehat{\boldsymbol{\theta}}$, on the basis of a suitable optimality criterion and submitting to a statistical analysis the residual errors $\widehat{\epsilon}_{m,j}$, i.e., the differences between the measured data and the corresponding optimal computed values, $\widehat{\epsilon}_{m,j} = d_{m,j} - y_{m,j}(\widehat{\boldsymbol{\theta}}) = d_{m,j} - \widehat{y}_{m,j}$.

If the model perfectly describes the experiments, the sample of residual errors does not contain systematic errors; thus, it must be compatible with the statistical distribution of the random experimental errors. All the systematic discrepancies eventually observed are attributed to the mathematical model, thus allowing a comparison between alternative models, since systematic errors can be decreased if a better model becomes available.

The optimality criterion used to compute the best form of the available model is based on the concept of likelihood, defined in Sect. 3.1 as the confidence $p(\mathrm{Ex} \mid \mathrm{Th})$ in obtaining the experimental result Ex if the theory Th is true. In the light of the above discussion, the likelihood can be intended as the probability of obtaining the residual errors, which depend on the experimental data and the model, through $\boldsymbol{y}$. Since for any component and any time instant the following is true:

$$p(\widehat{\epsilon}_{m,j}) = p(d_{m,j} \cap \mathrm{Th}) = p\left(\sigma_m^2, \widehat{\boldsymbol{\theta}}\right), \tag{3.13}$$

the likelihood of any residual error depends on the distribution function of the experimental errors and on the optimal values of the parameters. In order to find an explicit expression for this probability, some additional assumptions must be adopted, as discussed in the next section.

3.3 Maximum Likelihood and Least Squares Criteria

A few additional simplifying assumptions must be introduced, in order to obtain a workable expression for an objective function based on the concept of likelihood. First, it is assumed that the comparison between the experimental data and the optimized computed data produces random residual errors $\widehat{\epsilon}_{m,j}$ following a normal distribution with zero mean. Then, it is assumed that the experimental data are independent, i.e., all the covariances are zero, and the measurements of any component in any experiment have the same accuracy, i.e., $\sigma_{m,j}^2 = \sigma_m^2$.

Under these assumptions, the probability density function of observing the generic residual error can be expressed as

$$f_p(\widehat{\epsilon}_{m,j}) = \frac{\exp(-\frac{\widehat{\epsilon}_{m,j}^2}{2\sigma_m^2})}{\sqrt{2\pi}\sigma_m}, \tag{3.14}$$

where a similar expression is assumed to be valid for the errors $\epsilon_{m,j}$, i.e., for those errors obtained before the optimization procedure.

The objective function to be maximized is the total probability of these errors. Since the experimental data are assumed to be independent, the total probability can be obtained by multiplying the single probabilities. Thus, the maximum likelihood optimality criterion is obtained by exploiting the probability density function (3.14), i.e.,

$$\max_{\theta,\sigma_m} \left\{ \prod_{m=1}^{N_\mathrm{M}} \prod_{j=1}^{N_\mathrm{D}} \frac{\exp(-\frac{\widehat{\epsilon}_{m,j}^2}{2\sigma_m^2})}{\sqrt{2\pi}\sigma_m} \right\}. \tag{3.15}$$

This criterion can be recast in a more explicit form by considering that the maximum of (3.15) corresponds to the maximum of its logarithm; thus, the product of the probability densities (3.14) produces a sum of logarithms, i.e., the logarithmic maximum likelihood objective function to be maximized,

$$U_\mathrm{LML} = \sum_{m=1}^{N_\mathrm{M}} \sum_{j=1}^{N_\mathrm{D}} \left(-\frac{\epsilon_{m,j}^2}{2\sigma_m^2} - \frac{\ln(\sigma_m^2)}{2} - \frac{\ln(2\pi)}{2} \right). \tag{3.16}$$

By eliminating the constant terms it is possible to define the simplified logarithmic maximum likelihood function

$$U_\mathrm{LML}^* = \sum_{m=1}^{N_\mathrm{M}} \sum_{j=1}^{N_\mathrm{D}} \frac{\epsilon_{m,j}^2}{\sigma_m^2} + N_\mathrm{D} \sum_{m=1}^{N_\mathrm{M}} \ln(\sigma_m^2), \tag{3.17}$$

which has to be minimized to attain the maximum of U_LML.

Noticeably, when the variances σ_m^2 cannot be assigned a priori from the repetition of the experiments, expression (3.17) allows their computation. In fact, equating to zero the partial derivatives of the optimized value of U_LML^* with respect to σ_m^2

$$\frac{\partial \widehat{U}_\mathrm{LML}^*}{\partial \sigma_m^2} = -\sum_{j=1}^{N_\mathrm{D}} \frac{\widehat{\epsilon}_{m,j}^2}{(\sigma_m^2)^2} + \frac{N_\mathrm{D}}{\sigma_m^2} = 0, \quad m = 1, \ldots, N_\mathrm{M}, \tag{3.18}$$

yields, for any measured component,

$$\sigma_m^2 = \frac{1}{N_\mathrm{D}} \sum_{j=1}^{N_\mathrm{D}} \widehat{\epsilon}_{m,j}^2. \tag{3.19}$$

In conclusion, by substitution into (3.17), the *maximum likelihood criterion* can be expressed in the form

$$\min_{\boldsymbol{\theta}} \left\{ \sum_{m=1}^{N_\mathrm{M}} \ln \left(\sum_{j=1}^{N_\mathrm{D}} \widehat{\epsilon}_{m,j}^2 \right) \right\}, \tag{3.20}$$

so that the modified objective function

$$U_\mathrm{ML} = \sum_{m=1}^{N_\mathrm{M}} \ln \left(\sum_{j=1}^{N_\mathrm{D}} \epsilon_{m,j}^2 \right) \tag{3.21}$$

is usually named *maximum likelihood objective function.*

Instead, when the values of σ_m^2 can be assigned on the basis of repeated experimental measurements, the second term of the objective function in (3.17) becomes constant with respect to the parameters $\boldsymbol{\theta}$; therefore the function U_LML can be rearranged in the weighted least squares objective function

$$U_\mathrm{WLS} = \sum_{m=1}^{N_\mathrm{M}} \sum_{j=1}^{N_\mathrm{D}} \frac{\epsilon_{m,j}^2}{\sigma_m^2} = \sum_{m=1}^{N_\mathrm{M}} w_m \sum_{j=1}^{N_\mathrm{D}} \epsilon_{m,j}^2, \tag{3.22}$$

where the weights, chosen as $w_m = 1/\sigma_m^2$, are also useful in order to express in a dimensionless form the single terms of the sum in (3.22), which can have different physical dimensions. In this way the following *weighted least squares criterion* can be formulated:

$$\min_{\boldsymbol{\theta}} \left\{ \sum_{m=1}^{N_\mathrm{M}} w_m \sum_{j=1}^{N_\mathrm{D}} \widehat{\epsilon}_{m,j}^2 \right\}. \tag{3.23}$$

Sometimes, arbitrary values are assigned to the weights, in order to force the model to fit the experimental data; nevertheless, this practice has not a scientific value and can be accepted only when empirical models are used with practical purposes.

Finally, when all the measured variables are characterized by the same variance, i.e., $\sigma_m^2 = \sigma^2 \; \forall m = 1, \ldots, N_\mathrm{M}$, then the weighted least squares objective function can be rearranged in the simpler least squares objective function

$$U_\mathrm{LS} = \sum_{m=1}^{N_\mathrm{M}} \sum_{j=1}^{N_\mathrm{D}} \epsilon_{m,j}^2, \tag{3.24}$$

in which all the (equal) weights w_m can be neglected. On the basis of the objective function (3.24), the *least squares criterion* can be formulated as

$$\min_{\boldsymbol{\theta}} \left\{ \sum_{m=1}^{N_\mathrm{M}} \sum_{j=1}^{N_\mathrm{D}} \widehat{\epsilon}^{\,2}_{m,j} \right\}, \tag{3.25}$$

which is widely used, in particular for univariate problems ($N_\mathrm{M} = 1$).

For models in which the dependent variables are linear functions of the parameters, the solution to the above-mentioned optimization problems can be obtained in closed form when the least squares objective functions (3.22) and (3.24) are considered. However, in chemical kinetics, linear problems are encountered only in very simple cases, so that optimization techniques for nonlinear models must be considered.

The main methods for both linear and nonlinear optimization are presented in the following, with reference to the objective functions U_LS and U_WLS, since they allow for a more straightforward analysis. Hence, in the following, $U = U_\mathrm{LS}$ or $U = U_\mathrm{WLS}$; nevertheless, the analysis of nonlinear models, which is discussed in Sect. 3.5, can be extended with a little computational effort to the more rigorous maximum likelihood objective function (3.21).

3.4 Optimization for Models Linear in the Parameters

The general structure of models linear in the parameters is easily derived from (3.11) by considering the vector $\boldsymbol{y}$ of measured variables and by posing, for $m = 1, \ldots, N_\mathrm{M}$ and $p = 1, \ldots, N_\mathrm{P}$,

$$\frac{\partial \phi_m}{\partial \theta_p} = \frac{\partial y_m}{\partial \theta_p} = \varphi_{m,p}. \tag{3.26}$$

The following expression can be obtained:

$$\boldsymbol{y}(\boldsymbol{\theta}) = \boldsymbol{\phi}(t, t_0, \boldsymbol{x}_0, \boldsymbol{u}_{[t,t_0[}, \boldsymbol{\theta}) = \boldsymbol{\Phi}(t, t_0, \boldsymbol{x}_0, \boldsymbol{u}_{[t,t_0[})\boldsymbol{\theta}, \tag{3.27}$$

where

$$\boldsymbol{\Phi} = \begin{bmatrix} \varphi_{1,1} & \varphi_{1,2} & \cdots & \varphi_{1,N_\mathrm{P}} \\ \varphi_{2,1} & \varphi_{2,2} & \cdots & \varphi_{2,N_\mathrm{P}} \\ \vdots & \vdots & \vdots & \vdots \\ \varphi_{N_\mathrm{M},1} & \varphi_{N_\mathrm{M},2} & \cdots & \varphi_{N_\mathrm{M},N_\mathrm{P}} \end{bmatrix}$$

is a matrix whose elements $\varphi_{m,p}$ are defined in (3.26) and are called partial sensitivities or *sensitivity coefficients* [1]. Partial sensitivities measure the influence of the parameter θ_p on the variable y_m; in the case of linear-in-the-parameter models, these sensitivities are independent of $\boldsymbol{y}$ and of $\boldsymbol{\theta}$. In Sect. 3.6, the concept of partial sensitivities is extended to models nonlinear in the parameters.

The optimal value of the parameter vector, $\widehat{\boldsymbol{\theta}}$, is given by the solution of the following system of equations:

$$\frac{\partial U}{\partial \theta_p} = 0, \quad p = 1, \ldots, N_{\mathrm{P}}, \tag{3.28}$$

sometimes referred to as *normal equations*. When using the weighted least squares criterion, these equations can be set in an explicit form by using both (3.22), which gives

$$\frac{\partial U}{\partial \theta_p} = -2 \sum_{m=1}^{N_{\mathrm{M}}} w_m \sum_{j=1}^{N_{\mathrm{D}}} \left(\epsilon_{m,j} \left. \frac{\partial y_m}{\partial \theta_p} \right|_j \right) = -2 \sum_{m=1}^{N_{\mathrm{M}}} w_m \sum_{j=1}^{N_{\mathrm{D}}} \left[(d_{m,j} - y_{m,j}) \left. \frac{\partial y_m}{\partial \theta_p} \right|_j \right],$$

and model (3.27), which allows to compute $y_{m,j}$ and the partial sensitivities at each experimental time instant t_j

$$\left. \frac{\partial y_m}{\partial \theta_p} \right|_j = \frac{\partial}{\partial \theta_p} \left(\sum_{q=1}^{N_{\mathrm{P}}} \varphi_{m,q,j} \theta_q \right) = \varphi_{m,p,j}. \tag{3.29}$$

In conclusion, the following expression is obtained:

$$\frac{\partial U}{\partial \theta_p} = -2 \sum_{m=1}^{N_{\mathrm{M}}} w_m \sum_{j=1}^{N_{\mathrm{D}}} (d_{m,j} - y_{m,j}) \varphi_{m,p,j}, \quad p = 1, \ldots, N_{\mathrm{P}}, \tag{3.30}$$

i.e., a system of N_{P} linear equations in the unknowns $\boldsymbol{\theta}$, which can be written in vector notation as

$$\boldsymbol{\Gamma} \boldsymbol{\theta} = \boldsymbol{\psi}, \tag{3.31}$$

where $\boldsymbol{\Gamma}$ is the $(N_{\mathrm{P}} \times N_{\mathrm{P}})$ matrix of coefficients whose generic element is given by

$$\gamma_{p,q} = \sum_{m=1}^{N_{\mathrm{M}}} w_m \sum_{j=1}^{N_{\mathrm{D}}} \varphi_{m,p,j} \varphi_{m,q,j}, \tag{3.32}$$

and $\boldsymbol{\psi}$ is the $(N_{\mathrm{P}} \times 1)$ vector of known terms, whose generic element is given by

$$\psi_p = \sum_{m=1}^{N_{\mathrm{M}}} w_m \sum_{j=1}^{N_{\mathrm{D}}} d_{m,j} \varphi_{m,p,j}. \tag{3.33}$$

The solution of (3.31) can be obtained by multiplying both sides by the inverse of $\boldsymbol{\Gamma}$, namely

$$\widehat{\boldsymbol{\theta}} = \boldsymbol{\Gamma}^{-1} \boldsymbol{\psi}. \tag{3.34}$$

Of course, similar but simpler expressions are obtained when using the objective function (3.24) and, in particular, for univariate problems. It is worth noticing that the above expression coincides with that obtained by resorting to the so-called pseudoinverse of the matrix obtained by stacking the matrices $\boldsymbol{\Phi}$ computed at each experimental time instant [14].

3.5 Optimization for Models Nonlinear in the Parameters

For mathematical models that are nonlinear in the parameters and/or when the maximum likelihood objective function is used, the best estimates of the adjustable parameters must be obtained via iterative methods. These methods, initialized by a tentative first estimate of the parameter vector, $\boldsymbol{\theta}_0$, are based on a strategy for modifying the current estimate and include a convenient but arbitrary termination criterion to stop the algorithm. Thus, the results must be considered only approximate; moreover, if several local minima of the objective function are present, the attainment of the absolute minimum is not guaranteed since these methods usually provide a suboptimal solution, i.e., they converge to a local minimum depending on $\boldsymbol{\theta}_0$. A good solution to this problem is, when possible, to make use of a preliminary estimation procedure to produce a good initial estimate of the parameter values. If this is not possible, it can be useful to compare the results obtained using different initial estimates of the parameter vector.

According to [1], numerical optimization methods can be classified into:

- *zero-order methods*, based only on the values taken by the objective function at each step
- *first-order methods*, based on the values taken by the objective function and its gradient at each step; and
- *second-order methods*, based on the values taken by the objective function, its gradient, and its Hessian matrix at each step.

The zero-order methods are widely used when the model contains only one adjustable parameter, but, generally, they become ineffective in converging to the solution when the number of adjustable parameters is significantly larger than one. This is because these methods make use of heuristic strategies to move from the current estimate $\boldsymbol{\theta}_s$ to the updated value $\boldsymbol{\theta}_{s+1}$. On the contrary, the higher-order methods are characterized by the computation of a search direction and by the choice of a step length to take along this direction. In fact, the more effective zero-order methods also use the gradient concept, because a pseudo-optimal search direction may be determined on the basis of several values of the function evaluated around the current point [19]. In the following, the most popular first- and second-order optimization algorithms are briefly described.

3.5.1 *Steepest Descent Algorithm*

The steepest descent method, proposed by Cauchy in 1847 [8], is also known as gradient method. It is one of the oldest and simplest first-order algorithms for minimization of a real function defined in a vector space and represents the starting point for most of the more sophisticated optimization procedures.

The basic idea is that the gradient vector of the objective function, $\nabla U(\boldsymbol{\theta}) = [\partial U/\partial\theta_1 \cdots \partial U/\partial\theta_{N_\mathrm{P}}]$, represents the direction of faster increase of the function. Hence, the estimate at step $s+1$ can be computed via the recursive law

$$\boldsymbol{\theta}_{s+1} = \boldsymbol{\theta}_s - \kappa\nabla U(\boldsymbol{\theta}_s), \tag{3.35}$$

since, if the step size $\kappa > 0$ is small enough, then $U(\boldsymbol{\theta}_{s+1}) \leq U(\boldsymbol{\theta}_s)$.

Unfortunately, the gradient $\nabla U(\boldsymbol{\theta}_s)$ represents the best direction only in the neighborhood of $\boldsymbol{\theta}_s$; therefore, even though the convergence of the algorithm toward a local minimum can be proven [2], it may require a very large number of iterations; in particular, it converges very slowly in the neighborhood of the minimum where ∇U is small.

The optimal choice of the step size κ for any different optimization problem is not a trivial task; moreover, if the curvature of the objective function strongly changes along distinct directions, a faster convergence can be obtained by adopting a different value for the step size at each iteration. On the other hand, finding the optimal value of κ at each step can be very time-consuming; thus, for the sake of simplicity, in the following the step size is considered as a constant.

3.5.2 Newton–Raphson Algorithm

This method, first introduced by Isaac Newton and better formulated in the actual form by Joseph Raphson, is the simplest second-order algorithm. The basic idea is to use a quadratic approximation to the objective function around the initial parameter estimate and, then, to adjust the parameters in order to minimize the quadratic approximation until their values converge.

The function $U = U(\boldsymbol{\theta})$ can be approximated by the second-order Taylor series expansion at the point $\boldsymbol{\theta}_0$, i.e.,

$$U \simeq U_0 + \nabla U(\boldsymbol{\theta}_0)(\boldsymbol{\theta} - \boldsymbol{\theta}_0) + \frac{1}{2}(\boldsymbol{\theta} - \boldsymbol{\theta}_0)^\mathrm{T}\boldsymbol{H}(\boldsymbol{\theta}_0)(\boldsymbol{\theta} - \boldsymbol{\theta}_0), \tag{3.36}$$

where $\boldsymbol{H}(\boldsymbol{\theta})$ is the Hessian matrix, whose generic element $h_{p,q}$ is given by

$$h_{p,q} = \frac{\partial^2 U}{\partial\theta_p\partial\theta_q}. \tag{3.37}$$

By setting the derivative of the function (3.36) equal to zero yields

$$\frac{\partial U}{\partial\boldsymbol{\theta}} = \nabla U(\boldsymbol{\theta}_0) + \boldsymbol{H}(\boldsymbol{\theta}_0)(\boldsymbol{\theta} - \boldsymbol{\theta}_0) = 0, \tag{3.38}$$

so that the following value of the parameter vector can be found:

$$\boldsymbol{\theta} = \boldsymbol{\theta}_0 - \boldsymbol{H}^{-1}(\boldsymbol{\theta}_0)\nabla U(\boldsymbol{\theta}_0). \tag{3.39}$$

For nonquadratic objective functions, (3.39) can be used to obtain an iterative estimation law with fixed step size, $\kappa > 0$,

$$\boldsymbol{\theta}_{s+1} = \boldsymbol{\theta}_s - \kappa \boldsymbol{H}^{-1}(\boldsymbol{\theta}_s)\nabla U(\boldsymbol{\theta}_s). \tag{3.40}$$

The direction given by $-\boldsymbol{H}(\boldsymbol{\theta}_s)^{-1}\nabla U(\boldsymbol{\theta}_s)$ is a descent direction only when the Hessian matrix is positive definite. For this reason, the Newton–Raphson algorithm is less robust than the steepest descent method; hence, it does not guarantee the convergence toward a local minimum. On the other hand, when the Hessian matrix is positive definite, and in particular in a neighborhood of the minimum, the algorithm converges much faster than the first-order methods.

In some cases, the major disadvantage of second-order methods may be the programming effort required to derive explicit expressions for the Hessian elements, whose number increases as the square of the number of parameters. In Sect. 3.6, a simplified form of the Hessian matrix is derived by considering the particular form of the least-squares objective functions.

3.5.3 Levenberg–Marquardt Algorithm

If the Hessian matrix is badly conditioned, the computation of its inverse becomes numerically unstable and the recursive procedure (3.40) may be brought to divergence. To overcome this problem, several algorithms have been developed, in which the Hessian matrix is replaced by a suitable positive definite matrix $\boldsymbol{G}(\boldsymbol{\theta})$. The most important of these algorithms was firstly proposed by Kenneth Levenberg in 1944 [13] and then rediscovered and improved by Donald Marquardt in 1963 [15]. This method is more robust than the Newton–Raphson algorithm in terms of capability of finding a solution, even if the initial parameter estimate is far from the best estimate; on the other hand, it tends to be a bit slower when the starting point is very close to the minimum.

In detail, the iterative law is given by

$$\boldsymbol{\theta}_{s+1} = \boldsymbol{\theta}_s - \kappa \boldsymbol{G}^{-1}(\boldsymbol{\theta}_s)\nabla U(\boldsymbol{\theta}_s) \tag{3.41}$$

with a fixed step size $\kappa > 0$ and

$$\boldsymbol{G}(\boldsymbol{\theta}_s) = \boldsymbol{H}(\boldsymbol{\theta}_s) + \lambda \boldsymbol{I}, \tag{3.42}$$

where $\boldsymbol{I}$ is the identity matrix having the same dimensions of $\boldsymbol{H}$, and λ is a nonnegative damping factor.

The damping factor is of the utmost importance; in fact, several more or less heuristic arguments have been put forward for its best choice. Marquardt recommended to choose a sufficiently large initial value λ_0 [15]; indeed, by setting $\lambda = \lambda_0$, the matrix $\boldsymbol{G}(\boldsymbol{\theta}_s)$ is diagonal dominant, so that $\boldsymbol{G}(\boldsymbol{\theta}_s)\nabla U(\boldsymbol{\theta}_s) \simeq \lambda \nabla U(\boldsymbol{\theta}_s)$, and the Levenberg–Marquardt algorithm coincides with the steepest descent method. Subsequently, the damping factor can be reduced step by step by a factor $\nu > 1$ in such a way to tend to $\lambda = 0$ in the neighborhood of the minimum where the algorithm proceeds as the Newton–Raphson algorithm.

The Levenberg–Marquardt algorithm can be summarized in the following steps:

1. Choose a reasonable value of initial point $\boldsymbol{\theta}_0$, an initial value of the damping factor λ_0 and a factor $\nu > 1$.
2. Set $s = 0$, $\boldsymbol{\theta}_s = \boldsymbol{\theta}_0$, $\lambda = \lambda_0$.
3. Compute $U(\boldsymbol{\theta}_s)$.
4. Compute an update $\boldsymbol{\theta}_{s+1}$ of the parameter vector via (3.41).
5. Compute $U(\boldsymbol{\theta}_{s+1})$ and compare it with $U(\boldsymbol{\theta}_s)$; then:
   (i) If $U(\boldsymbol{\theta}_{s+1}) \le U(\boldsymbol{\theta}_s)$, the algorithm proceeds in the right direction; if the termination criterion is not yet fulfilled, then set $s = s + 1$, $\boldsymbol{\theta}_s = \boldsymbol{\theta}_{s+1}$, and $\lambda = \lambda/\nu$.
   (ii) If $U(\boldsymbol{\theta}_{s+1}) \ge U(\boldsymbol{\theta}_s)$, the algorithm proceeds in the wrong direction; thus preserve the current values of $\boldsymbol{\theta}$ and U, set $\lambda = \nu\lambda$, and return to step 4.
6. Repeat steps from 3 to 5 until a suitably chosen termination check is verified.

3.6 Implicit Models

Systems encountered in chemical kinetics can be very often represented by implicit models in the form (3.10), for which the explicit solution can be obtained only in a few simple cases. Since for these models, the variables $\boldsymbol{y}$, to be compared with the experimental data, are not available, it is also impossible to directly compute both the gradient and the Hessian matrix of the objective function.

Let us first consider the gradient vector of the weighted least square objective function (3.22), whose generic component has the form

$$\frac{\partial U}{\partial \theta_p} = -2 \sum_{m=1}^{N_M} w_m \sum_{j=1}^{N_D} \epsilon_{m,j} \left. \frac{\partial y_m}{\partial \theta_p} \right|_j = -2 \sum_{m=1}^{N_M} w_m \sum_{j=1}^{N_D} \epsilon_{m,j} \varphi_{m,p,j}, \tag{3.43}$$

where the partial sensitivities $\varphi_{m,p,j}$ can be defined as in (3.26) for models linear in the parameters. However, it must be remarked that, for models nonlinear in the parameters, the partial sensitivities are functions of both $\boldsymbol{x}$ and $\boldsymbol{\theta}$.

In order to compute the partial sensitivities for implicit models of the form (3.10) the following procedure must be used [1, 5]. First, it must be observed that all the components of (3.10) must be considered, i.e., not only those related to the measured variables y_m, since in nonlinear models each measured component y_m can depend on the complete set of dependent variables.

By differentiating both sides of the hth component of (3.10) with respect to θ_p, the following equality is obtained:

$$\frac{\partial \dot{x}_h}{\partial \theta_p} = \frac{\partial f_h}{\partial \theta_p} + \sum_{i=1}^{N_C} \frac{\partial f_h}{\partial x_i} \frac{\partial x_i}{\partial \theta_p}. \tag{3.44}$$

Interchanging the order of differentiation and introducing the sensitivities yields

$$\dot{\varphi}_{h,p} = \frac{d}{dt}\left(\frac{\partial x_h}{\partial \theta_p}\right) = \frac{\partial f_h}{\partial \theta_p} + \sum_{i=1}^{N_C} \frac{\partial f_h}{\partial x_i}\varphi_{i,p}, \tag{3.45}$$

where $\partial f_h/\partial \theta_p$ and $\partial f_h/\partial x_i$ can be easily obtained by simple differentiation of the model equations. The differential equations (3.45) are called sensitivity equations [1]; in order to obtain the sensitivity coefficients, they must be integrated, together with the model equations (3.10). In the case under study, these equations must be integrated with the initial conditions $\varphi_{h,p}(t_0) = 0\ \forall h, p$, since the initial values of concentration do not depend on the parameters.

This procedure allows computing the gradient ∇U as detailed in (3.43). Nevertheless, in the case of implicit models, it is impossible to directly compute the Hessian matrix in the form defined in (3.37), since its generic term $h_{p,q}$, which is easily obtained by differentiating (3.43),

$$\frac{\partial^2 U}{\partial \theta_p \partial \theta_q} = 2\sum_{m=1}^{N_M} w_m \sum_{j=1}^{N_D} \left.\frac{\partial y_m}{\partial \theta_p}\right|_j \left.\frac{\partial y_m}{\partial \theta_q}\right|_j - 2\sum_{m=1}^{N_M} w_m \sum_{j=1}^{N_D} \epsilon_{m,j} \left.\frac{\partial^2 y_m}{\partial \theta_p \partial \theta_q}\right|_j, \tag{3.46}$$

contains the second derivatives of y_m.

In the case under study, the particular form of the adopted objective functions allows one to overcome this difficulty by introducing a simplified form of the Hessian matrix. In fact, by assuming that the errors $\epsilon_{m,j}$ are small, the second term on the right-hand side of (3.46) can be neglected, and the Hessian matrix can be approximated by the first-order term, which only contains the first derivatives. This assumption can be always done in a neighborhood of the minimum, where these errors tend to the residuals. In conclusion, the form

$$\frac{\partial^2 U}{\partial \theta_p \partial \theta_q} = 2\sum_{m=1}^{N_M} w_m \sum_{j=1}^{N_D} \varphi_{m,p,j}\varphi_{m,q,j} \tag{3.47}$$

can be used in the optimization procedure. The simplified form (3.47) can be used in the case of explicit models as well, whenever the computation of the second derivatives is considered too onerous; in particular, it works well together with the Levenberg–Marquardt algorithm, which uses the Hessian matrix only in a neighborhood of the optimum.

3.7 Statistical Analysis of the Results

Once the best estimates of the adjustable parameters have been computed, an analysis of the results allows one to evaluate the quality of the correspondence between experimental data and mathematical model and to identify the best model among the available alternatives. This analysis consists of different steps, mainly based on the principles of statistics, and, for the sake of simplicity, is briefly discussed hereafter with reference to simple linear-in-the-parameter models. The results may be extended with some caution to models nonlinear in the parameters, by considering the linearization of the original model around the optimal point.

First, the hypothesis that the residuals $\widehat{\epsilon}_{m,j}$ represent a sample of the universe of the experimental errors can be tested via different methods in order to single out the presence of systematic errors deriving from the inadequacy of the mathematical model. In particular, it is possible to test whether, for any measured component m, the mean of the residuals $\widehat{\epsilon}_{m,j}$ (for $j = 1, \ldots, N_\mathrm{D}$) is significantly different from zero (which is the expected value) and whether, for any measured component, the corrected residual variance $\hat{s}^2_{\mathrm{R},m}$ is significantly different from the universe variance σ^2_m, which can be computed by resorting to repeated measurements, as shown by (3.8).

Interestingly, any detected nonrandom behavior of the residuals may provide useful suggestions to improve the adopted model. In the case of chemical kinetics, by plotting the residuals against the reaction time, it is possible to single out the need for including a new reaction and/or a new intermediate component in the kinetic scheme.

Since the best estimate $\widehat{\boldsymbol{\theta}}$ must be considered as a random variable, a third step of this analysis allows one to evaluate the accuracy of the estimate. In fact, when the estimation of the kinetic parameters is characterized by a large variance $\sigma^2_{\theta,p}$, the relevant reaction is not supported by the experimental data, and thus, the experimental campaign must be extended, and/or the model must be changed.

For example, let us consider the simple scalar linear model

$$y = \varphi\theta, \tag{3.48}$$

which leads to the relationship

$$\widehat{\theta} = \frac{\widehat{y}_j}{\varphi_j} = \frac{d_j}{\varphi_j} - \frac{\widehat{\epsilon}_j}{\varphi_j} \tag{3.49}$$

between the best estimate and each residual. If all the residuals are extracted from the same universe, i.e., if they all have the same variance σ^2, (3.49) yields

$$\sigma^2_{\widehat{\theta}} = 2\frac{\sigma^2}{\sum_{j=1}^{N_\mathrm{D}} \varphi_j^2} = 2\frac{\sigma^2}{\partial^2 U/\partial\theta^2}, \tag{3.50}$$

which can be generalized for a multivariable model in the form

$$\boldsymbol{V} = 2\sigma^2 \boldsymbol{H}^{-1}, \tag{3.51}$$

where $\boldsymbol{V}$ is the matrix of the covariances between the parameters.

Finally, a quantitative comparison between alternative models must be considered. On the basis of the previously discussed results in Sects. 3.1 to 3.3, this comparison should be based on the ratio of likelihoods (3.6), each of them computed as a product of probabilities. In practice, the likelihood is assumed to be inversely proportional to the corrected variance of the residuals that, for simple scalar problems, is given by

$$\hat{s}_{\mathrm{R}}^2 = \frac{1}{N_{\mathrm{D}} - N_{\mathrm{P}}} \sum_{j=1}^{N_{\mathrm{D}}} \widehat{\epsilon}_j^2 = \frac{\widehat{U}_{\mathrm{LS}}}{N_{\mathrm{D}} - N_{\mathrm{P}}}. \tag{3.52}$$

This relationship can be extended to multivariate problems, even if, in this case, it is questionable how to subdivide the degrees of freedom $N_{\mathrm{D}} - N_{\mathrm{P}}$ among the different components. It should be also underlined that $\hat{s}_{\mathrm{R}}^2 \to \infty$ as $N_{\mathrm{D}} - N_{\mathrm{P}} \to 0$; this clearly shows that using a too large number of parameters, or even resorting to a collocation polynomial, is not a proper scientific procedure.

3.8 Case Study: Identification of Reduced Kinetic Models

In this section, the phenol–formaldehyde reactive system is considered as an example of identification of reduced kinetic models. The kinetic model containing 13 components and 89 reactions, developed in Sect. 2.4 to study the production of 1,3,5-methylolphenol, is too detailed and complex for control and monitoring purposes. Thus, in this section this model is referred to as *detailed model*, while four reduced kinetic models, based on lumped components and reactions, are developed.

First, the detailed model is used to simulate the behavior of the real system, and a set of simulated isothermal experimental data is generated including the total heat released by reaction. Then, these data are used to estimate the kinetic parameters of the reduced models and the heats of reaction of the lumped reactions. Finally, the reduced kinetic models are tested in a validation procedure which simulates the operation of a batch reactor and allows one to identify the best reduced model.

3.8.1 *Reduced Models*

Many techniques aiming at reducing the complexity of kinetic models [16, 21] and at identifying the parameters of the relevant reduced models have been proposed [5, 6]. Among the most successful approaches, elimination of reactions and species [4, 9, 20, 22] (assisted by the use of genetic algorithms, sensitivity analysis, quasi steady-state approximation, or computational singular perturbations) have been widely applied to describe complex reaction mechanisms by means of simplified mathematical models. Moreover, lumping procedures have also shown major advantages in the modeling of reactive systems [18].

According to the approach described in [11], two alternative reduced kinetic models are proposed here to describe the phenol–formaldehyde reaction network introduced in Sect. 2.4. This approach includes, first, the selection of a general class of simplified models and, among this class, the selection of a number of different candidate models to be compared. Three main classes of model structure can be considered [7, 11]: black-box model, parameter white-box model, and hybrid grey-box model. A hybrid approach is pursued here, because the reduced structure of the model is derived on the basis of the available knowledge of the system.

Two different reduced kinetic models have been considered, both involving four lumped chemical species and three and four lumped reactions, respectively. These models describe the time evolution of the following components:

- Phenol, here denoted as reactant A. Since the second reactant, formaldehyde, is fed to the reactor in large excess, its concentration can be assumed as a constant during the reaction; thus, it does not explicitly appear in the rate expressions and has not been considered in the reduced kinetic models.
- the sum of the substituted monomers, i.e., of the two mono- and the two di-methylolphenols, denoted as reaction intermediate I and given by

$$\mathrm{I} = \sum_{i=1}^{4} \mathrm{MPh}_i . \tag{3.53}$$

- 1,3,5-methylolphenol, here denoted as desired product P.
- The sum of all di-phenols and poli-phenols, here denoted as undesired product W and given by

$$\mathrm{W} = \sum_{i=0}^{4} \mathrm{DPh}_i + \mathrm{PPh}. \tag{3.54}$$

The first reduced kinetic model is a series of three consecutive reactions

$$\mathrm{A} \xrightarrow{k_{c,1}} \mathrm{I} \xrightarrow{k_{c,2}} \mathrm{P} \xrightarrow{k_{c,3}} \mathrm{W}, \tag{3.55}$$

whereas the second one includes an additional parallel reaction

$$\begin{aligned} &\mathrm{A} \xrightarrow{k_{c,1}} \mathrm{I} \xrightarrow{k_{c,2}} \mathrm{P} \xrightarrow{k_{c,3}} \mathrm{W}, \\ &\mathrm{A} + \mathrm{I} \xrightarrow{k_{c,4}} \mathrm{W}. \end{aligned} \tag{3.56}$$

For each kinetic scheme, the relevant mathematical model is given by the mass balances written for the four species involved. In detail, for the scheme (3.55), the mass balance yields

$$\begin{aligned} \dot{y}_1 &= -R_1, \\ \dot{y}_2 &= R_1 - R_2, \\ \dot{y}_3 &= R_2 - R_3, \end{aligned} \tag{3.57}$$

where y_1, y_2, and y_3 are the computed concentrations of reactant A, intermediate I, and product P, respectively, and R_l ($l = 1, 2, 3$) are the reaction rates. It has not

Table 3.1 Reaction rate expressions for the reduced kinetic models

	Kinetic model (3.57)		Kinetic model (3.58)	
	First order	Second order	First order	Second order
$R_1 =$	$k_{c1}y_1$	$k_{c1}y_1^2$	$k_{c1}y_1$	$k_{c1}y_1^2$
$R_2 =$	$k_{c2}y_2$	$k_{c2}y_2^2$	$k_{c2}y_2$	$k_{c2}y_2^2$
$R_3 =$	$k_{c3}y_3$	$k_{c3}y_3^2$	$k_{c3}y_3$	$k_{c3}y_3^2$
$R_4 =$	–	–	$k_{c4}y_2$	$k_{c4}y_1y_2$

been necessary to introduce the mass balance on the final product W because its concentration linearly depends on the concentrations of A, I, and P and can be more easily obtained through the conservation of the total mass. According to (3.57), the mass balances for the scheme (3.56) are given by

$$\begin{aligned}\dot{y}_1 &= -R_1 - R_4,\\ \dot{y}_2 &= R_1 - R_2 - R_4,\\ \dot{y}_3 &= R_2 - R_3.\end{aligned} \tag{3.58}$$

For both models, first-order and second-order kinetics have been considered, as reported in Table 3.1. Moreover, the Arrhenius law

$$k_{cl}(T_r) = k_{0l}\exp\left(-\frac{E_{al}}{\mathcal{R}T_r}\right), \quad l = 1, \ldots, N_L, \tag{3.59}$$

has been introduced to describe the temperature dependence of the rate constants in terms of preexponential factors and activation energies, where $N_L = 3$ for the first kinetic model, and $N_L = 4$ for the second kinetic model.

The computed specific thermal power, i.e., the heat released per unit volume and unit time by the chemical reactions (3.55) and (3.56), is given by

$$y_{\dot{q}} = -\sum_{l=1}^{N_L} \Delta H_{Rl} R_l. \tag{3.60}$$

It is worth noticing that, in the presence of lumped compounds and lumped reactions, the molar heats of reaction ΔH_{Rl} are not known and may vary significantly with temperature.

3.8.2 Generation of Data for Identification

The detailed kinetic model has been used to simulate the behavior of the reactive system in MATLAB/SIMULINK© by performing $N_Z = 9$ isothermal runs at different temperatures T_z, equally spaced by 5°C from $T_1 = 60$°C to $T_9 = 100$°C. For each run, the specific thermal power $\dot{q}$ and the concentrations of the 13 measured components C, defined in Sect. 2.4.1, have been stored at 71 different times t_j ($j = 1, \dots, 71$) up to 15 hours. Smaller time intervals have been adopted for the first two hours, since the reaction is faster in the initial phases. Finally, in order to simulate measurement errors, Gaussian white noise has been added to the computed values; this noise has zero mean and standard deviations of 10 mol m^{-3} and 0.1 kJ m^{-3} s^{-1} for concentrations and heat, respectively.

Then, the concentration data have been lumped into the three components A, I, and P defined in Sect. 3.8.1. In conclusion, $9 \times 71 = 639$ simulated measurements of $\dot{q}$ and $3 \times 9 \times 71 = 1917$ measurements of concentrations have been obtained. It is worth remarking that the above-described simulated measurements are easily available in a real context. In fact, as discussed in Sect. 2.6, the concentrations can be measured by drawing a sample of reacting mixture and analyzing it offline, while the heat released by the reactions can be obtained via calorimetric methods.

3.8.3 Estimating the Kinetic Parameters

The identification problem described above is somewhat different and more complex as compared to the implicit isothermal problem described in Sect. 3.6, where N_P rate constants are estimated from data measured at constant temperature. In fact, in this case, for each reduced model, the unknown parameters are ($l = 1, \dots, N_\text{L}$):

- the preexponential factors, k_{0l}
- the activation energies $E_{\text{a}l}$; and
- the molar enthalpy changes $\Delta H_{\text{R}l}$.

In order to solve the identification problem, the method for implicit nonlinear models discussed in the previous sections must be conveniently modified. In particular, the estimation problem has been divided into two subproblems by estimating, in different procedures, first the kinetic parameters and then the molar heats of reaction.

The more usual procedure for estimating k_{0l} and $E_{\text{a}l}$ from experimental data taken at different temperatures consists in considering N_Z distinct isothermal problems and estimating the relevant values of the rate constants; then, from these data and the relationship (3.59) it is possible to estimate k_{0l} and $E_{\text{a}l}$. Nevertheless, since the law describing the temperature dependence of the rate constants is known, it is possible to estimate directly k_{0l} and $E_{\text{a}l}$. To deal with 9 different isothermal runs, it is only necessary to repeat the integration 9 times for each computational step of the objective function; in other words, the dimension N_Z of the data can be eliminated by posing in series the 9 sets of data.

The major drawback of this procedure consists in the covariance between k_{0l} and $E_{\text{a}l}$, which can be explained by a simple geometrical argument. In the linear relationship between $\ln(k_{\text{c}l})$ and $E_{\text{a}l}$, shown by (3.59), k_{0l} represents the intercept at $1/T_\text{r} = 0$, i.e., the limit for $T_\text{r} \to \infty$ of $k_{\text{c}l}$, which is usually very far from the (usually small) experimental range of temperature; thus, any small uncertainty in $E_{\text{a}l}$,

Table 3.2 Kinetic parameters for the reduced models[a]

	Kinetic model (3.57)		Kinetic model (3.58)	
	First order	Second order	First order	Second order
k_{01}	$(2.44 \pm 1.56)10^7$	$(2.40 \pm 0.76)10^5$	$(0.20 \pm 1.85)10^7$	$(3.19 \pm 0.35)10^5$
E_{a1}	77.88 ± 2.96	87.31 ± 0.67	70.51 ± 2.96	88.13 ± 0.35
k_{02}	$(1.21 \pm 0.60)10^4$	$(4.32 \pm 1.11)10^3$	$(3.80 \pm 1.76)10^4$	$(4.73 \pm 1.88)10^3$
E_{a2}	58.27 ± 0.83	77.94 ± 0.96	61.55 ± 1.82	78.41 ± 1.07
k_{03}	$(1.23 \pm 0.77)10^2$	$(1.69 \pm 0.54)10$	$(1.50 \pm 0.28)10^3$	$(2.32 \pm 1.29)10^2$
E_{a3}	43.78 ± 2.25	59.79 ± 2.18	51.56 ± 3.58	68.02 ± 3.31
k_{04}	–	–	$(4.57 \pm 0.51)10^4$	$(2.88 \pm 0.42)10$
E_{a4}	–	–	249.58 ± 5.81	71.01 ± 4.10

[a]The activation energies E_{al} are expressed in [kJ mol^{-1}], the preexponential factors k_{0l} are in [s^{-1}] for models characterized by first-order kinetics and in [m^3 mol^{-1} s^{-1}] for models characterized by second-order kinetics

due to experimental errors, produces a very large uncertainty in k_{0l}. As a consequence, it may be very difficult to identify the best pair (k_{0l}, E_{al}) among several almost equivalent alternatives. This drawback can be eliminated (or at least reduced), as recommended in [6], by substituting k_{0l} with a new parameter k_{cl}^{o},

$$k_{cl}^{o} = k_{0l} \exp\left(-\frac{E_{al}}{\mathcal{R}T_r^o}\right), \tag{3.61}$$

where T_r^o is a suitable reference temperature; thus, (3.59) becomes

$$k_{cl} = k_{cl}^{o} \exp\left[\frac{E_{al}}{\mathcal{R}}\left(\frac{1}{T_r^o} - \frac{1}{T}\right)\right]. \tag{3.62}$$

Since T_r^o is located inside the experimental range, k_{cl}^{o} results to be almost independent of E_{al}.

The best parameter values have been computed by minimizing a least squares objective function, based on the concentration of the three measured species A, I, and P, namely

$$U_C = \sum_{m=1}^{3}\sum_{z=1}^{9}\sum_{j=1}^{71}(C_{m,z,j} - y_{m,z,j})^2, \tag{3.63}$$

and by resorting to the Levenberg–Marquardt algorithm reviewed in Sect. 3.5.3. The following values for the algorithm parameters have been adopted: $\kappa = 10^{-1}$, $\lambda_0 = 10^3$, and $\nu = 10$. Since a reliable first estimate has not been available a priori, the iterative routine has been executed for 256 different initial values of the parameter vector. The optimal values obtained for the kinetic parameters k_{0l} and E_{al} are reported in Table 3.2.

In general, the activation energies are estimated with a fair accuracy, whereas the estimates of the preexponential factors suffer from the problem discussed above because the experimental temperature range is rather limited. Nevertheless, it must be underlined that this problem does not affect very much the estimates of the rate constants inside the examined experimental range of temperature.

Moreover, the parameters of the kinetic models (3.58) are estimated more accurately, with the exception of E_{a3}, whereas no significant difference can be observed between the two models (3.58). However, it must be remarked that the accuracy of the estimates does not affect the quality of the model fitting to the experimental data; to this purpose, a validation of the different reduced models is presented in Sect. 3.8.5.

3.8.4 Estimating the Heats of Reaction

Once the kinetic parameters have been estimated, (3.60) becomes linear in the unknown parameters $\Delta H_{\mathrm{R}l}$. Therefore, the errors between the total heat of reaction, computed via the detailed model, and the total heat, computed via each reduced model, can be minimized by resorting to the least squares solution of a linear regression problem, discussed in Sect. 3.4. The molar heats of reaction, included in the vector of parameters

$$\boldsymbol{\theta} = [\Delta H_{\mathrm{R}1} \quad \dots \quad \Delta H_{\mathrm{R}N_{\mathrm{L}}}]^{\mathrm{T}}, \tag{3.64}$$

can be identified by considering the following objective function:

$$U_{\dot{q}} = \sum_{z=1}^{9} \sum_{j=1}^{71} (\dot{q}_{z,j} - y_{\dot{q}_{z,j}})^2, \tag{3.65}$$

where $\dot{q}_{z,j}$ and $y_{\dot{q}_{z,j}}$ are the experimental and computed thermal powers at temperature T_z and time t_j, respectively.

Unfortunately, the molar enthalpy changes, which are almost independent of temperature for true reactions, are biased by the effect of lumping; hence, it is necessary to identify a distinct value for each temperature and each reaction. Thus, the best values of $\Delta H_{\mathrm{R}l}$ for each temperature T_z have been identified by considering 9 different vectors of parameters, i.e., $\boldsymbol{\theta}_z = [\Delta H_{\mathrm{R}1}(T_z) \dots \Delta H_{\mathrm{R}N_{\mathrm{L}}}(T_z)]^{\mathrm{T}}$. Then, the effect of temperature has been described by an empirical polynomial law, i.e., for each reaction

$$\Delta H_{\mathrm{R}l}(T_{\mathrm{r}}) = \alpha_{0,l} + \alpha_{1,l} T_{\mathrm{r}} + \alpha_{2,l} T_{\mathrm{r}}^2 + \alpha_{3,l} T_{\mathrm{r}}^3, \tag{3.66}$$

and the coefficients $\alpha_{i,l}$ in (3.66) have been determined by interpolating the values of $\Delta H_{\mathrm{R}l}(T_z)$ obtained. Figure 3.1 shows an example of these results.

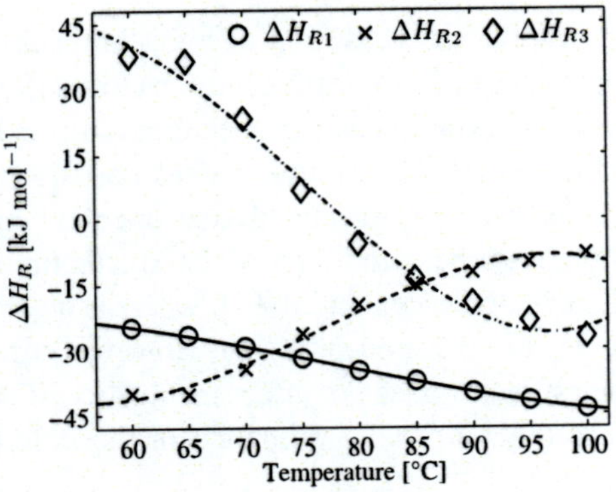

Fig. 3.1 Interpolating the molar enthalpy changes for model (3.57) with first-order kinetics

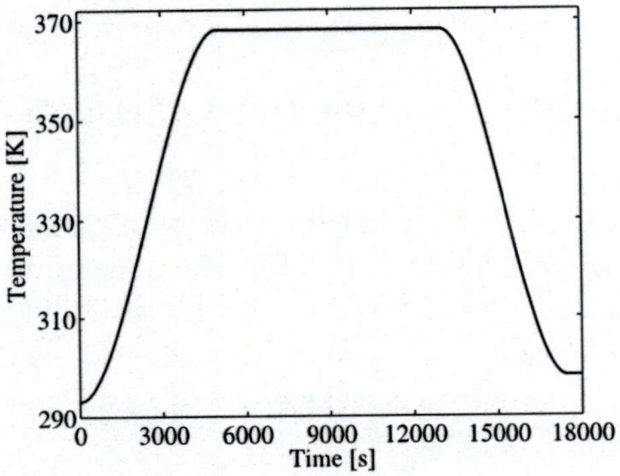

Fig. 3.2 Temperature profile

3.8.5 Validation of the Reduced Models

The quality of the reduced kinetic models, as compared to the detailed model, has been evaluated in simulation by comparing their ability in tracking a few assigned temperature profiles. For the sake of simplicity, only the results obtained for the temperature profile shown in Fig. 3.2 are presented. The test profile considered consists of three steps: heating of reactants up to a set-point temperature, reaction phase at constant temperature, and cooling down to ambient conditions.

The models are compared in Table 3.3, in terms of the root mean squared errors for the concentrations

$$\sigma_C = \sqrt{\frac{1}{3N_\mathrm{D}} \sum_{m=1}^{3} \sum_{j=1}^{N_\mathrm{D}} (C_{m,j} - y_{m,j})^2} \tag{3.67}$$

and for the specific thermal power

$$\sigma_{\dot{q}} = \sqrt{\frac{1}{N_\mathrm{D}} \sum_{j=1}^{N_\mathrm{D}} (\dot{q}_j - y_{\dot{q}\,j})^2}. \tag{3.68}$$

Table 3.3 Root mean squared error of concentration and specific thermal power

	Kinetic model (3.57)		Kinetic model (3.58)	
	First order	Second order	First order	Second order
σ_C [mol m^{-3}]	0.560	0.180	0.502	0.170
$\sigma_{\dot{q}}$ [kJ m^{-3} s^{-1}]	0.648	3.157	0.750	0.131

Table 3.4 Percent phenol conversion X, asymptotic values of concentration computed with the detailed and the reduced kinetic models for phenol, $C_{A,\infty}$, and trimethylolphenol, $C_{P,\infty}$, and relevant percent errors relative to the detailed model used as reference (indicated by the superscript ^{o})

		Kinetic model (3.57)		Kinetic model (3.58)	
	Detailed	First order	Second order	First order	Second order
X, percent	82.22	89.93	81.10	87.28	81.94
$C_{A,\infty}$ [kmol]	746.61	422.97	793.90	543.30	758.64
$\Delta C_A/C^o_{A,\infty}$, percent		−43.35	6.33	−27.23	1.61
$C_{P,\infty}$ [kmol]	1146.93	1001.35	1191.17	1027.17	1151.29
$\Delta C_P/C^o_{P,\infty}$, percent		−12.69	3.86	−10.44	0.38

In Table 3.4, the comparison is performed in terms of phenol conversion and of the asymptotic values of phenol ($C_{A,\infty}$) and trimethylolphenol ($C_{P,\infty}$) concentrations. Thus, whereas the data in Table 3.3 account for the accuracy in reconstructing the concentrations and the specific thermal power, the data in Table 3.4 are related to the accuracy in predicting the final values of concentration, which are of more practical interest.

Moreover, the results are graphically shown in Figs. 3.3–3.6. In detail, Fig. 3.3 shows the results obtained with the kinetic model (3.57) with first-order kinetics. The fitting of concentrations (left) is rather poor; in particular, the asymptotic values at the largest reaction times are not correctly estimated. This reduced model underestimates the final product concentration and overestimates the final conversion of phenol by more than 7 percent, which corresponds to an error of more than 43 percent on the phenol concentration and of about 13 percent on product concentration (Table 3.4). A better fitting is obtained for the specific thermal power (Fig. 3.3, right).

When a second-order kinetics is introduced in this kinetic model, the fitting of concentrations is markedly improved, as shown in Fig. 3.4 (left) and in Table 3.4. The error in the final concentrations is strongly reduced to about 6 and 4 percent, respectively; nevertheless, a very poor fitting of the specific thermal power is obtained (Fig. 3.4, right). This result may be explained by observing that the lumping procedure of the reaction network strongly biases the physical meaning of the estimated heats of reaction, whose fitting is obtained in the terms of an empirical function.

A parallel reaction is introduced in the kinetic model (3.58). When the results obtained with first-order kinetics, shown in Fig. 3.5, are compared with the corresponding first-order kinetic model with reactions in series (Fig. 3.3), a slight im-

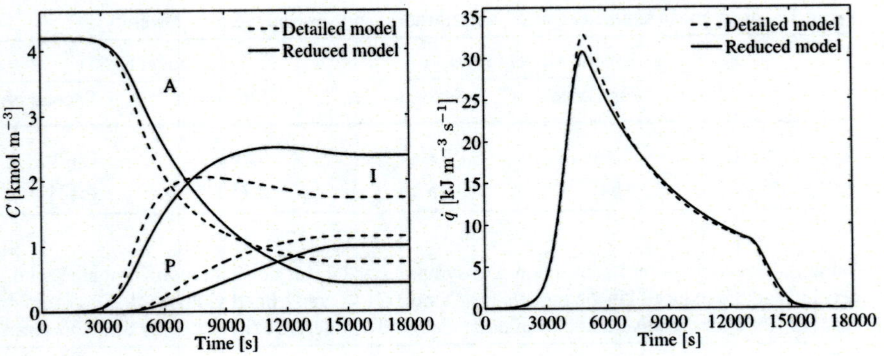

Fig. 3.3 Concentrations and specific thermal power computed via model (3.57) with first-order kinetics

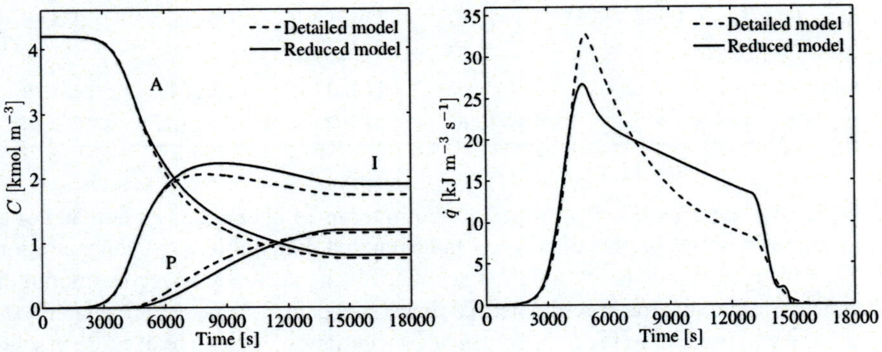

Fig. 3.4 Concentrations and specific thermal power computed via model (3.57) with second-order kinetics

provement of the fitting of concentrations, as well as a poorer fitting of the specific thermal power, can be noticed.

On the contrary, the introduction of a second-order kinetics in this model produces the best results for all the measured variables, as shown in Fig. 3.6; in particular, the errors on concentrations of phenol and product are reduced to about 1.6 and 0.4 percent, respectively, while the errors on the specific thermal power are very small.

In synthesis, the analysis of the results suggests the following remarks:

- the best match, in terms of both concentration and specific thermal power, is obtained via model (3.58) with second-order kinetics
- both second-order models are characterized by good performance in terms of accuracy of concentration estimation
- despite its good performance in terms of concentration estimation, model (3.57) with second-order kinetics performs badly in terms of specific thermal power accuracy; and

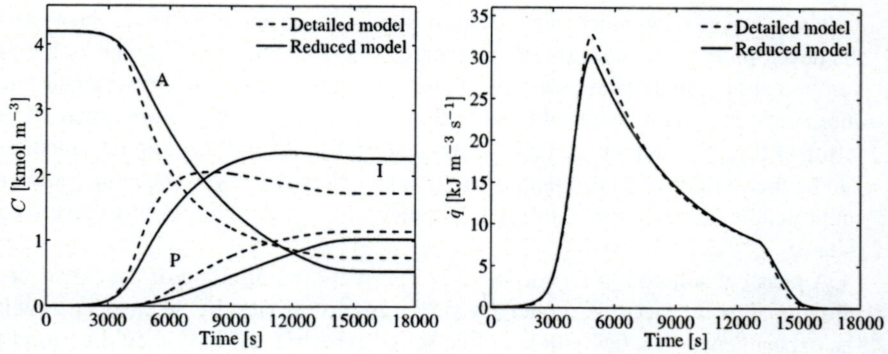

Fig. 3.5 Concentrations and specific thermal power computed via model (3.58) with first-order kinetics

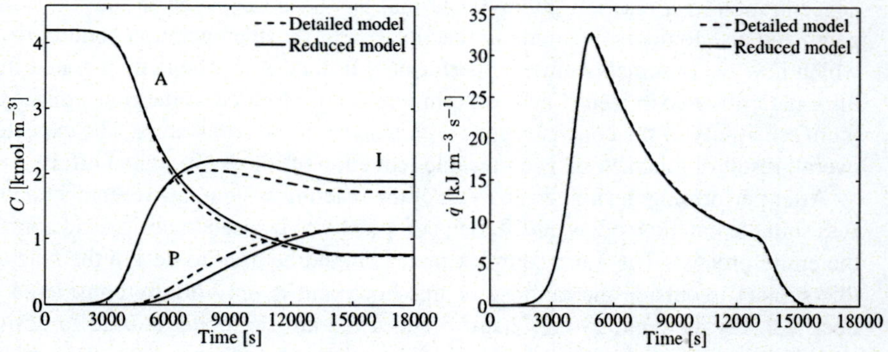

Fig. 3.6 Concentrations and specific thermal power computed via model (3.58) with second-order kinetics

- the first-order models present similar and rather poor results: model (3.58) achieves a better accuracy for the concentrations estimates, while model (3.57) achieves a better accuracy for the specific thermal power.

3.9 Conclusions

The methods discussed in this chapter represent a rigorous approach to the identification of mathematical models and the estimation of the relevant adjustable parameters. In particular, these methods may allow one to obtain significant results in the kinetic analysis of chemical reacting systems, as can be argued from the example reported in Sect. 3.8 and several examples in the literature [12]. Nevertheless, at least in some cases, a practicing engineer might consider these methods to be more fitted for a study performed with a scientific purpose rather than for the use in industry, where the need of obtaining acceptable results in times consistent with the development of the industrial process may suggest a more straightforward procedure.

The personal experience of the authors allows us to conclude that, in most cases, the development of a suitable comprehensive kinetic model is not limited by the development of a mathematical model but by the limitations of the experimental measurements. In fact, in chemical kinetics the experiments are very often expensive, in terms of time and money, and not all the kinetically relevant reaction intermediates can be measurable and even defined in their chemical structure. Thus, the testing of more detailed kinetic models may be hindered by the availability of experimental data.

A possible solution to this problem is a correct programming of the experimental campaign. In fact, the model should be improved step by step, together with the experiments. In a few words, a first set of experimental data should be used to test simple kinetic models; then, on the basis of the inaccuracies of these models, a new, more exhaustive and accurate experimental campaign should be planned and performed. This method may allow one to point out the necessity of measuring the concentration of a reaction intermediate that has been neglected in the first campaign and also to define the optimal reaction times, i.e., the operating conditions in which new measurements must be performed. In fact, it can be demonstrated that information is maximized if new experiments are performed at the times of maximum sensitivity of the concentrations with respect to the parameters. The expected overall result of this method is noticeable reduction of the experimental effort.

After performing the kinetic analysis of the reacting system, the researchers possess suitable kinetic models of different complexity to be used to design and control the entire process. The more complex model should be used to design the reactor; this subject is outside the purpose of this book and is only briefly considered in Sect. 7.4. On the contrary, in Chaps. 5 and 6 the kinetic model is used to design adaptive model-based control and fault diagnosis schemes for a class of reactions taking place in batch reactors.

References

1. Y.A. Bard. *Nonlinear Parameter Estimation*. Academic Press, New York, 1974.
2. R.D. Bartusiak, C. Georgakis, and M.J. Reilly. Nonlinear feedforward/feedback control structures designed by reference synthesis. *Chemical Engineering Science*, 44:1837–1851, 1989.
3. T. Bayes. An essay towards solving a problem in the doctrine of chances. *Philosophical Transactions of the Royal Society of London*, 53:370–418, 1763.
4. B. Bhattacharjee, D.A. Schwer, P.I. Barton, and W.H. Green Jr. Optimally-reduced kinetic models: reaction elimination in large-scale kinetic mechanisms. *Combustion and Flame*, 135:191–208, 2003.
5. L.T. Biegler and J.J. Damiano. Nonlinear parameter estimation: a case study comparison. *AIChE Journal*, 32(1):29–45, 1986.
6. P. Bilardello, X. Joulia, J.M. Le Lann, H. Delmas, and B. Koehret. A general strategy for parameter estimation in differential-algebraic systems. *Computers and Chemical Engineering*, 17(5/6):517–525, 1993.
7. D. Bonvin. Optimal operation of batch reactors—a personal view. *Journal of Process Control*, 8(5/6):355–368, 1998.
8. A. Cauchy. Méthodes générales pour la résolution des systémes d'équations simultanées. *C. R. Acad. Sci. Paris*, 25:536–538, 1847.

9. K. Edwards, T.F. Edgar, and V.I. Manousiouthakis. Kinetic model reduction using genetic algorithms. *Computers and Chemical Engineering*, 22:239–246, 1998.
10. R.A. Fisher. *Statistical Methods, Experimental Design and Scientific Inference*. Oxford University Press, Oxford, 1990.
11. M.A. Henson and D.E. Seborg. *Nonlinear Process Control*. Prentice Hall, Upper Saddle River, 1997.
12. D.M. Himmelblau and K. B. Bishoff. *Process Analysis and Simulation*. Wiley, New York, 1968.
13. K. Levenberg. A method for the solution of certain non-linear problems in least squares. *The Quarterly of Applied Mathematics*, 2:164–168, 1944.
14. L. Ljung. *System Identification. Theory for the User*. Prentice Hall, Upper Saddle River, 1999.
15. D. Marquardt. An algorithm for least squares estimation of nonlinear parameters. *SIAM Journal on Applied Mathematics*, 11:431–441, 1963.
16. W. Marquardt. Nonlinear model reduction for optimization based control of transient chemical processes. *AIChE Symposium Series 326*, 98:12–42, 2001.
17. K. Popper. *The Logic of Scientific Discovery*. Hutchinson, London, 1959.
18. E. Ranzi, M. Dente, A. Goldaniga, G. Bozzano, and T. Faravelli. Lumping procedures in detailed kinetic modeling of gasification, pyrolysis, partial oxidation and combustion of hydrocarbon mixtures. *Progress in Energy and Combustion Science*, 27:99–139, 2001.
19. S.S. Rao. *Engineering Optimization: Theory and Practice, 4th Edition*. Wiley, Hoboken, 2009.
20. W. Su and H. Huang. Development and calibration of a reduced chemical kinetic model of n-heptane for HCCI engine combustion. *Fuel*, 84:1029–1040, 2005.
21. F. Tjärnström and L. Ljung. L_2 model reduction and variance reduction. *Automatica*, 38:1517–1530, 2002.
22. P.F. Tupper. Adaptive model reduction for chemical kinetics. *BIT Numerical Mathematics*, 42:447–465, 2002.

Chapter 4
Thermal Stability

List of Principal Symbols

B	dimensionless number defined in (4.36)
c	mass heat capacity [J kg^{-1} K^{-1}]
$\mathcal{C}$	dimensionless concentration
C_{A}	concentration of reactant A [mol m^{-3}]
E_{a}	activation energy [J mol^{-1}]
h	incremental step
ΔH_{R}	molar enthalpy change of reaction [J mol^{-1}]
k_0	preexponential factor [s^{-1}]
q_{E}	dimensionless rate of heat exchange
q_{R}	dimensionless rate of heat production by reaction
$\mathcal{R}$	universal gas constant [J mol^{-1} K^{-1}]
s	normalized objective sensitivity
S	heat transfer area [m^2]
Se	dimensionless Semenov number
t	time [s]
t_{E}	characteristic time of heat exchange [s]
t_{R}	characteristic reaction time [s]
T	temperature [K]
$\mathcal{T}$	dimensionless temperature
U	overall heat transfer coefficient [J m^{-2} K^{-1} s^{-1}]
V	volume [m^3]

Greek Symbols

θ	generic model parameter
Λ	dimensionless group defined in (4.7)
ρ	density [kg m^{-3}]
τ	dimensionless time
τ_{b}	dimensionless batch time
τ_{I}	dimensionless induction time
τ_{M}	dimensionless time to maximum reaction rate

F. Caccavale et al., *Control and Monitoring of Chemical Batch Reactors*,
Advances in Industrial Control,
DOI 10.1007/978-0-85729-195-0_4,

Φ dimensionless group defined in (4.8)
Ω dimensionless group defined in (4.6)

Subscripts and Superscripts

ad	adiabatic conditions
c	critical value
j	jacket
ma	maximum allowable value
max	maximum value
r	reactor
0	initial value
o	reference value

4.1 Runaway in Chemical Batch Reactors

Several reactions commonly encountered in the chemical industry are characterized by the ability to accelerate even without external intervention. This autocatalytic behavior follows two possible mechanisms of positive feedback (sketched in Fig. 4.1) having a thermal and a chemical nature, respectively. The thermal mechanism can be observed during exothermic reactions, when the heat generated by reaction is much larger than the heat leaving the system in thermal sinks (artificial cooling, heat losses through the reactor walls, parallel endothermic reactions, phase transitions phenomena, etc.). This causes an increase in the reactor temperature that, according to the Arrhenius term (2.2), exponentially increases the reaction rate; thus, the excess heat generated causes a further increase in temperature, then again in the reaction rate, and so on. On the other side, the chemical mechanism can be observed in branching chain reactions when the rate of formation of highly reactive intermediates (the so-called *chain carriers*, often radical species) largely exceeds their rate of disappearance in chemical sinks (chain closing reactions, chemical quenching on solid surfaces, etc.). In both cases the reaction accelerates and, if not properly controlled, eventually goes out of control (runaway from set point conditions). These two mechanisms can also occur simultaneously, as commonly observed in gas-phase combustion reactions.

Of the two mechanisms discussed above, thermal runaway is by far the most common cause of safety problems in chemical batch reactors, given the ability of the system to largely exceed the desired reactor temperature and, hence, the normal operative pressure with high risk of explosion. It has been estimated that an important fraction of the chemical reactions executed daily in the chemical industry has heat effects large enough to eventually cause reactor thermal runaway [16] and that ineffective temperature control has been the cause of many incidents involving batch reactors [4, 6].

Hence, when dealing with exothermic reactions, it is of utmost importance to determine in which conditions the batch reactor operation can be considered intrinsically safe. This is the subject of the present chapter, where guidelines are given for

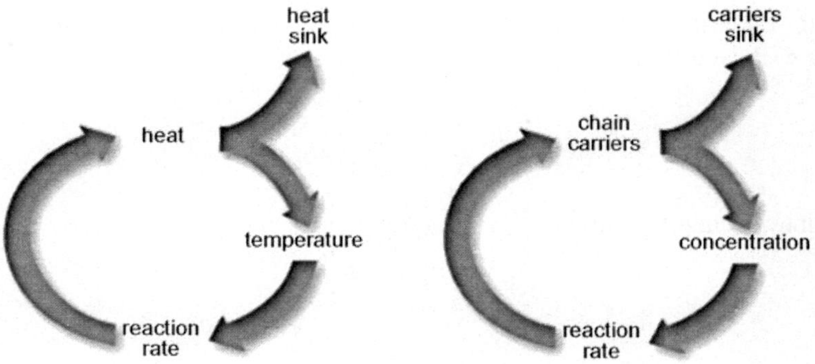

Fig. 4.1 Positive feedback in thermal and chemical runaway

the location of the so-called runaway boundaries, i.e., the critical conditions that separate the safe operation from the thermal explosion. The thermal stability of batch reactors is discussed here with reference to noncontrolled systems under adiabatic and isoperibolic (at constant coolant temperature) conditions, while guidelines for the setting up of effective control laws are then given in Chap. 5. The analysis is carried out by using a dimensionless form of the mathematical model of the reactor and under the simplifying assumptions of perfect mixing and negligible thermal inertia of the reactor walls, which is often satisfied in very large vessels.

4.2 Dimensionless Mathematical Model

Consider the exothermic first-order reaction A $\longrightarrow$ B taking place batchwise at reactor temperature T_r and coolant temperature T_j. The mathematical model describing the system is given by the mass balance on reactant A and the energy balance in the reactor:

$$\dot{C}_A = -k_0 \exp\left(-\frac{E_a}{RT_r}\right) C_A, \tag{4.1}$$

$$\rho_r c_r V_r \dot{T}_r = (-\Delta H_R) k_0 \exp\left(-\frac{E_a}{RT_r}\right) C_A V_r - US(T_r - T_j), \tag{4.2}$$

with initial conditions

$$C_A(0) = C_{A0}, \qquad T_r(0) = T_{r0}. \tag{4.3}$$

Here, the usually small variations of the parameters $(\rho_r, c_r, U, \Delta H_R)$ and of the concentration C_A with respect to temperature is neglected.

For a more systematic approach, the model equations can be rewritten in terms of dimensionless time, concentration, and temperature defined, respectively, as

$$\tau = \frac{t}{t_R}, \qquad \mathcal{C} = \frac{C_A}{C_{A0}}, \qquad \mathcal{T}_r = \frac{T_r}{T_r^o}, \tag{4.4}$$

where T_r^o is an arbitrary reference temperature, and t_R is the characteristic reaction time evaluated at T_r^o and given by

$$t_R = \frac{1}{k_0 \exp(-\frac{E_a}{\mathcal{R}T_r^o})}. \tag{4.5}$$

By introducing the dimensionless groups

$$\Omega = \frac{E_a}{\mathcal{R}T_r^o}, \tag{4.6}$$

$$\Lambda = \frac{(-\Delta H_R)C_{A0}}{\rho_r c_r T_r^o}, \tag{4.7}$$

$$\Phi = \frac{US}{V_r \rho_r c_{vr} k_0 \exp(-\Omega)}, \tag{4.8}$$

$$\mathcal{T}_j = \frac{T_j}{T_r^o}, \tag{4.9}$$

$$\mathcal{T}_{r0} = \frac{T_{r0}}{T_r^o}, \tag{4.10}$$

the model equations become

$$\frac{d\mathcal{C}}{d\tau} = -\exp\left[-\Omega\left(\frac{1-\mathcal{T}_r}{\mathcal{T}_r}\right)\right]\mathcal{C}, \tag{4.11}$$

$$\frac{d\mathcal{T}_r}{d\tau} = \Lambda \exp\left[-\Omega\left(\frac{1-\mathcal{T}_r}{\mathcal{T}_r}\right)\right]\mathcal{C} - \Phi(\mathcal{T}_r - \mathcal{T}_j), \tag{4.12}$$

with initial conditions

$$\mathcal{C}(0) = 1, \qquad \mathcal{T}_r(0) = \mathcal{T}_{r0}. \tag{4.13}$$

The dimensionless numbers introduced above can provide, in the light of their physical meaning, some preliminary information about the system behavior:

- Ω is the so-called Arrhenius number, i.e., the dimensionless activation energy that quantifies the local sensitivity of the reaction rate with respect to deviations of the reactor temperature from the reference value. The higher the value of Ω, the faster the temperature increase during a thermal explosion.
- Λ is the maximum possible increase of the dimensionless temperature in the reactor and is calculated as the maximum temperature increase

$$\Delta T_{ad} = \frac{(-\Delta H_R)C_{A0}}{\rho_r c_r}, \tag{4.14}$$

corresponding to complete conversion of reactant A in adiabatic condition, divided by the reference temperature. This parameter quantifies the ability of the reactive system to self-heating.

- Φ is the ratio of the characteristic reaction time t_R defined in (4.5) to the characteristic time of heat exchange t_E defined as

$$t_E = \frac{\rho_r c_r V_r}{US}. \tag{4.15}$$

According to the definition (4.8), operation with very low Φ results in higher reactor temperatures and higher risk of runaway, because of ineffective reactor cooling. On the contrary, in the limit $\Phi \to \infty$, the reactor temperature approaches T_j during the entire reaction cycle.

The definition of the dimensionless numbers given here is slightly different from what can be found in the classical literature on this subject and is based on the more practical idea to keep different effects separated, i.e., the merely physico-chemical properties of the reactive mixture (Ω and Λ), the performance of the cooling system (Φ), the coolant temperature ($\mathcal{T}_j$), and the initial temperature ($\mathcal{T}_{r0}$). In this way, in design and scale-up problems the optimal range of Φ can be determined in order to prevent runaway conditions, while keeping the remaining parameters unchanged. On the other hand, when considering a given reactor (i.e., a fixed value of Φ), safe operative conditions can be determined in terms of temperatures $\mathcal{T}_{r0}$ and $\mathcal{T}_j$ and of the initial reactant concentration (accounted for in Λ).

By integrating (4.11) and (4.12), the concentration and temperature profiles in the reactor can be obtained, and conditions leading to reactor runaway can be investigated. Numerical solutions are required even for the simple kinetic scheme adopted here because of the nonlinear nature of the Arrhenius term.

Different criteria have been introduced in the past decades to individuate runaway boundaries in batchwise operated reactors. Most of them can be used to ensure a safe batch operation only when the reaction kinetics is fully known and the hypothesis of perfect mixing is satisfied. These criteria also strongly depend on the mode of operation with respect to heat exchange. Excluding isothermal conditions, the following modes of operation can be considered:

- *Adiabatic*: heat exchange with cooling media and heat losses through reactor walls are absent.
- *Isoperibolic*: the system exchanges heat with a cooling medium kept at constant temperature.
- *Temperature-controlled*: the coolant temperature and/or flow rate is adjusted, in order to obtain a desired temperature profile in the reactor.

In this chapter, the reactor dynamics under adiabatic and isoperibolic conditions is analyzed, while the temperature-controlled case is addressed in Chap. 5. It must be pointed out that these conditions can be easily realized in laboratory investigations, e.g., in reaction calorimetry, but represent mere ideality at the industrial scale. Nevertheless, this classification is useful to recognize the main paths leading to runaway without the burden of a more complex mathematical approach.

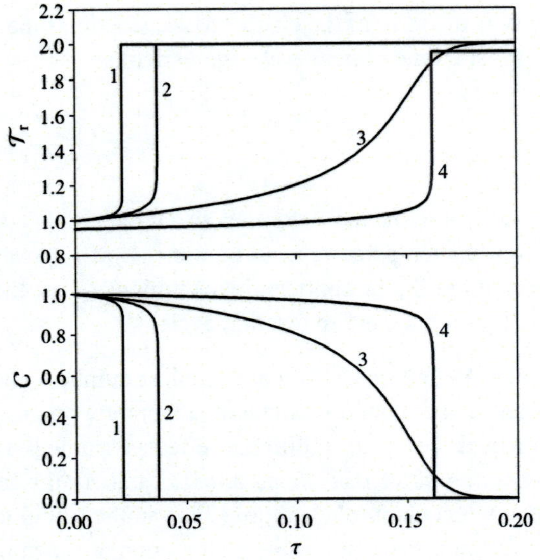

Fig. 4.2 Temperature and concentration profiles in adiabatic conditions at $\Omega = 50$ (curve *1*), 30 (curves *2* and *4*) and 10 (curve *3*). All curves are obtained with $\Lambda = 1$ and $\mathcal{T}_{r0} = 1$ except curve *4*, which is obtained with $\Lambda = 1$ and $\mathcal{T}_{r0} = 0.95$

4.3 Adiabatic Reactor

Under the assumption of adiabatic conditions, all the heat of reaction is converted into sensible heat, and, consequently, the final temperature of an adiabatic reaction is the maximum temperature that can be reached in a given reactive system. In this case, by imposing $\Phi = 0$ in (4.12), the system dynamics only depends on the values of Λ, Ω, and $\mathcal{T}_{r0}$.

In general, three main reaction stages can be recognized in the temperature and concentration profiles of an adiabatic batch cycle where thermal explosion occurs, as shown by curves 1 and 2 in Fig. 4.2 obtained with $\mathcal{T}_0 = 1$ and $\Lambda = 1$. In the first stage (induction), the reaction is extremely slow, and, consequently, heat accumulated in the system produces a very low temperature increase. After an induction time τ_I, a second stage takes place (ignition or thermal explosion), characterized by an extremely high reaction rate and an abrupt temperature increase that reaches its final value $\mathcal{T}_{r0} + \Lambda$ (i.e., the adiabatic dimensionless temperature), while reactant concentration rapidly goes to zero. In the third stage, reaction rate is zero, due to the full consumption of reactant.

Ignition becomes much less abrupt when decreasing Ω (curve 3), due to the lower sensitivity of the reaction rate to temperature. In this case, which is perhaps more representative of chemical reactors, whereas the previous cases better describe explosive systems, the same final values of temperature and concentration are reached through a smoother profile.

The effect of $\mathcal{T}_{r0}$ can be observed with reference to curve 4 in Fig. 4.2: when starting with a reactor temperature $\mathcal{T}_{r0} = 0.95$, the thermal explosion takes more time to occur, while the contrary happens when $\mathcal{T}_{r0} > 1$ (case not shown in the figure).

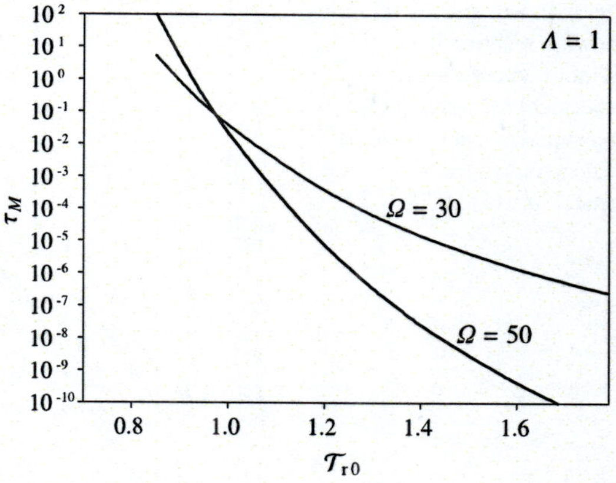

Fig. 4.3 Time needed to reach the maximum reaction rate as a function of $\mathcal{T}_{r0}$

To quantify the ignition delay, the time τ_M needed by the system to reach the maximum reaction rate can be considered (this also corresponds to the time at which the inflection point on the temperature–time profile occurs). The values of τ_M are reported in Fig. 4.3, for two different values of Ω, as a function of $\mathcal{T}_{r0}$. The effect of $\mathcal{T}_{r0}$ in accelerating the reaction proves to be very strong since, even for a small change in the dimensionless initial temperature, the time to the maximum reaction rate experiences a change of some orders of magnitude. As expected, the sensitivity to temperature is higher for higher values of Ω; this causes a crossing of the curves as shown in Fig. 4.3.

For reactions characterized by high values of Ω, the onset of a thermal explosion can be controlled by adjusting the batch time, τ_b: the explosion occurs if $\tau_b > \tau_I$ and does not in the opposite case. Hence, the reactor performance shows a typical on–off behavior, being characterized by either complete or negligible reactant conversion. It follows that reactions of interest must be carried out under explosion conditions, provided that the reactor vessel withstands the final internal pressure and the thermal shock caused by the sudden temperature increase. A similar map can be drawn with reference to undesired secondary reactions and, in this respect, operative parameters must be adjusted in order to avoid the ignition stage within the batch length.

4.4 Isoperibolic Reactor

The assumption of constant wall temperature is often more realistic for chemical reactors than the adiabatic case. In this respect, starting from the pioneering theory of thermal explosions developed by Semenov at the beginning of the last century [8], significant advances have been made by the related scientific literature with approaches that can be roughly classified as geometric and sensitivity-based, as described in detail in the following.

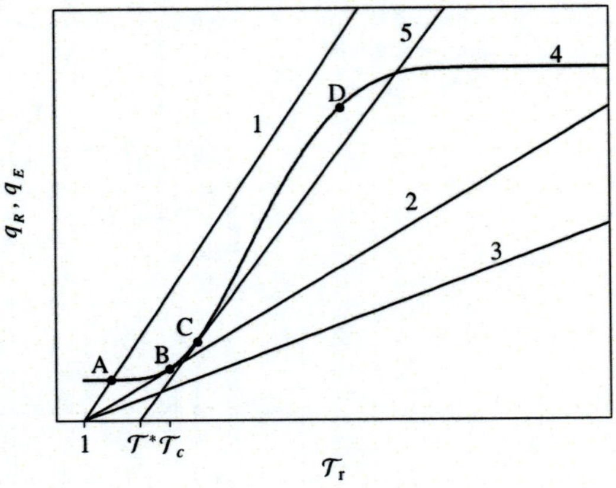

Fig. 4.4 Semenov diagram showing reaction heat sigmoid (curve *4*) and subcritical (*1*), critical (*2*), and supercritical (*3*) lines of heat exchange for $\mathcal{T}_j = 1$ and critical line (*5*) for $\mathcal{T}_j = \mathcal{T}^*$

4.4.1 The Semenov Theory

The theory of Semenov [8] was originally derived for zero-order reactions and can be applied whenever the reactant conversion is negligible and the reaction is very temperature-sensitive. It proves, however, to be too conservative in other cases. Nevertheless, it is resumed here since it gives a fundamental view of the dynamics of a thermal explosion and correct results in many practical cases.

By renaming the two terms on the right-hand side of (4.12) as q_R and q_E, which represent the rate of heat production by reaction and of heat exchange with the cooling medium, respectively, the heat balance in the batch reactor can be rewritten as

$$\frac{d\mathcal{T}_r}{d\tau} = q_R - q_E \tag{4.16}$$

with

$$q_R = \Lambda \exp\left[-\Omega\left(\frac{1-\mathcal{T}_r}{\mathcal{T}_r}\right)\right]\mathcal{C}, \tag{4.17}$$

$$q_E = \Phi(\mathcal{T}_r - \mathcal{T}_j). \tag{4.18}$$

Under the assumption $\mathcal{C} = 1$ at each time τ, the system evolves toward steady-state conditions that can be located graphically on the Semenov diagram of Fig. 4.4 as the intersections of the curves q_R and q_E; this condition implies, indeed, that $d\mathcal{T}_r/d\tau = 0$ in (4.16). For the sake of simplicity, let us first assume that $\mathcal{T}_{r0} = \mathcal{T}_j = 1$. When q_E is given by line 1 with slope Φ_1, the steady-state condition is given by point A, characterized by a low operating temperature. Point A is an attractor since its temperature is spontaneously restored after any small perturbation of the system and, consequently, in these conditions thermal explosion does not occur.

On decreasing Φ, the slope of q_E decreases until the critical (unstable) point B is reached where q_E is tangent to q_R (line 2); the corresponding critical temperature in B is denoted by $\mathcal{T}_c$. The critical slope Φ_2 is the lowest value that still leads to a bounded steady-state temperature. By further decreasing of Φ (line 3, $\Phi_3 < \Phi_2$),

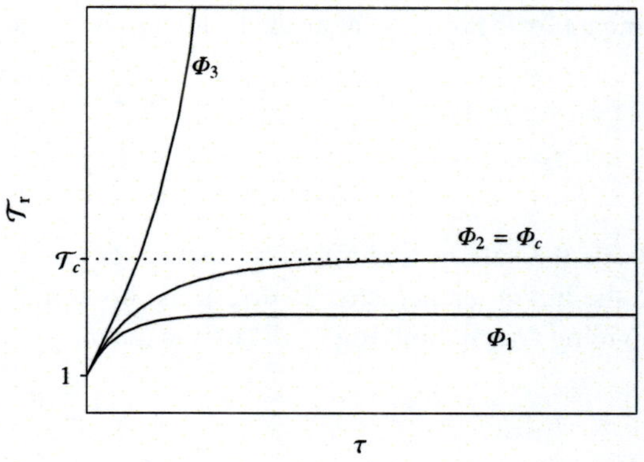

Fig. 4.5 Temperature profiles in subcritical (Φ_1), critical (Φ_2), and supercritical (Φ_3) conditions of heat exchange

there is no intersection in the low-temperature range, and the only steady-state solution can be found on the horizontal asymptote of q_R, so that the system evolves toward this point with a thermal explosion.

In Fig. 4.5, temperature profiles are reported in subcritical, critical, and supercritical conditions. Supercritical solutions of the simplified mathematical model proposed by Semenov are, however, purely theoretical since the assumption of negligible reactant conversion becomes very unrealistic. As an example, in the worst case where $\Phi \to 0$, the theory predicts an infinitely increasing temperature in the reactor.

For any fixed set of parameters, the critical point for runaway depends also on the temperature of the cooling medium. For example, if $\mathcal{T}_j = \mathcal{T}^* > 1$, the heat exchange is represented in Fig. 4.4 by line 5, and the new critical point is C. Clearly, the new critical Φ number has increased with respect to the previous case since a lower cooling time is required to balance the decreased driving force of heat exchange, $\mathcal{T}_r - \mathcal{T}_j$.

The critical point can be determined by imposing the tangency conditions [17], i.e.,

$$q_R = q_E, \tag{4.19}$$

$$\frac{dq_R}{d\mathcal{T}_r} = \frac{dq_E}{d\mathcal{T}_r}. \tag{4.20}$$

Two temperature values satisfy these conditions, namely

$$\mathcal{T}_c = \frac{\Omega}{2}\left(1 - \sqrt{1 - \frac{4\mathcal{T}_j}{\Omega}}\right), \tag{4.21}$$

$$\mathcal{T}_c' = \frac{\Omega}{2}\left(1 + \sqrt{1 - \frac{4\mathcal{T}_j}{\Omega}}\right). \tag{4.22}$$

The first value corresponds to tangency in point B and is the critical temperature defined above, while the second one corresponds to tangency in the high temperature region (surroundings of point D) and is discarded. The value of $\mathcal{T}_c$ does not directly depend on the particular choice of the initial temperature $\mathcal{T}_{r0}$, but the existence of tangent lines holds as far as the following necessary conditions are also satisfied:

$$\mathcal{T}_{r0} < \mathcal{T}_c, \tag{4.23}$$

$$\mathcal{T}_j < \frac{\Omega}{4}. \tag{4.24}$$

While the latter is true for most practical cases, the former poses an important limit to the initial temperature. In detail, the limit of $\mathcal{T}_c$ as $\Omega \to \infty$ can be obtained by dividing and multiplying (4.21) by the quantity

$$1+\sqrt{1-\frac{4\mathcal{T}_j}{\Omega}} \tag{4.25}$$

and reads

$$\lim_{\Omega\to\infty} \mathcal{T}_c = \lim_{\Omega\to\infty} \frac{\Omega}{2} \frac{\frac{4\mathcal{T}_j}{\Omega}}{1+\sqrt{1-\frac{4\mathcal{T}_j}{\Omega}}} = \mathcal{T}_j. \tag{4.26}$$

Consequently, condition (4.23) requires that, for very large Ω, the difference between the coolant and the initial temperature must be strictly positive.

By introducing the Semenov number

$$Se = \frac{\Lambda\Omega}{\Phi} \tag{4.27}$$

in (4.20), in critical conditions the following is obtained:

$$Se_c = \frac{1}{\mathcal{T}_c^2} \exp\left[\Omega\left(\frac{1-\mathcal{T}_c}{\mathcal{T}_c}\right)\right]. \tag{4.28}$$

Since, for a given reactive system, Ω and Λ are constant quantities, this corresponds to the critical cooling number given by

$$\Phi_c = \Omega\Lambda Se_c. \tag{4.29}$$

In Fig. 4.6, the critical Semenov number given by (4.28) is reported *versus* the coolant temperature $\mathcal{T}_j$ for different values of Ω. Operation is safe when $Se < Se_c$, provided that condition (4.23) is also satisfied. The critical Semenov number decreases as $\mathcal{T}_j$ increases, with a slope that increases with Ω, and, in the limit $\Omega \to \infty$, it reduces to the vertical line $\mathcal{T}_j = 1$; in this case, operation is safe whenever $\mathcal{T}_{r0} < \mathcal{T}_j < 1$. Moreover, all curves nearly cross each other at $\mathcal{T}_j = 1$, where their value can be approximated by

$$Se_c = \exp(-1) = 0.368, \tag{4.30}$$

which represents the classical simplified result of the Semenov theory on thermal explosion that defines the runaway boundaries for large Ω values.

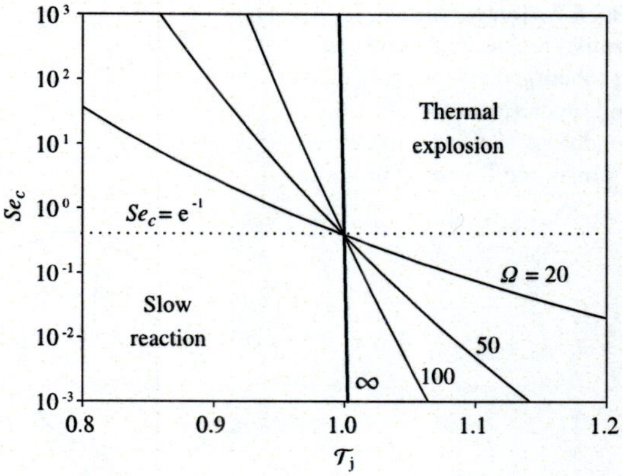

Fig. 4.6 Critical Semenov number as a function of $\mathcal{T}_j$ for different values of Ω

4.4.2 Geometry-based Runaway Criteria

Less conservative criteria for runaway can be found by removing the assumption of negligible reactant consumption. Along this line, a class of runaway criteria has been devised by linking the reactor behavior to suitable geometric features of the temperature-time history.

Thomas and Bowes [11] observed that, under runaway conditions, two inflection points exist before the maximum in the temperature-time plane, while they are missing in slow reaction conditions. Critical conditions are, then, defined as those where the inflection points first appear before the temperature maximum, i.e.,

$$\frac{d\mathcal{T}_r}{d\tau} > 0, \tag{4.31}$$

$$\frac{d^2\mathcal{T}_r}{d\tau^2} = 0, \tag{4.32}$$

$$\frac{d^3\mathcal{T}_r}{d\tau^3} = 0. \tag{4.33}$$

These conditions are satisfied when the second derivative of the temperature profile with respect to time has a maximum and this maximum value is zero. In Fig. 4.7, these requirements are satisfied by curve 2.

Adler and Enig [1] applied the same criterion to the temperature-conversion plane, where slightly less conservative runaway boundaries can be computed. To this goal, by dividing (4.12) by (4.11), the derivative of temperature with respect to concentration is obtained:

$$\frac{d\mathcal{T}_r}{d\mathcal{C}} = -\Lambda + \frac{\Phi}{\mathcal{C}} \exp\left[\Omega\left(\frac{1-\mathcal{T}_r}{\mathcal{T}_r}\right)\right](\mathcal{T}_r - \mathcal{T}_j) \tag{4.34}$$

with initial condition

$$\mathcal{T}_r(1) = \mathcal{T}_{r0}. \tag{4.35}$$

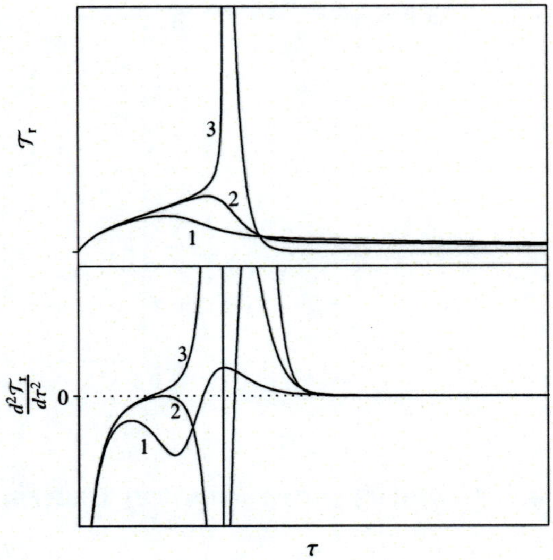

Fig. 4.7 Temperature profiles and second derivative in subcritical (*1*), critical (*2*), and supercritical (*3*) conditions according to the Thomas and Bowes criterion

The runaway boundaries are then determined by imposing the appearance of two coinciding inflection points, i.e., the same conditions as in (4.31), (4.32), and (4.33), but with the independent variable τ replaced by $\mathcal{C}$. Varma et al. [14] showed that, when the dimensionless group

$$B = \Lambda\Omega \tag{4.36}$$

is large, the criterion of Adler and Enig gives the same results of the Semenov criterion; this happens since, in this case, the thermal explosion occurs in the very early stages of the batch cycle, when reactant conversion is still negligible. When B decreases, however, the Adler and Enig criterion predicts a larger Se_c and appears, therefore, less conservative.

Still with reference to the temperature–concentration profile, van Welsenaere and Froment [13] proposed a criterion based on the locus of the temperature maxima that was originally derived for homogeneous tubular reactors but whose validity for batch reactors was also proved. The criterion is discussed here with reference to Fig. 4.8, where the temperature–concentration profiles in a batch reactor are reported for $Se = 0.470$, $\Omega = 40$, $\mathcal{T}_{r0} = \mathcal{T}_j = 1$, and different values of Λ in the range 0.2–1.16. The maxima of the $\mathcal{T}_r(\mathcal{C})$ curves (continuous lines) define a new curve (dashed line), which has itself a maximum with respect to $\mathcal{T}_r$. According to the criterion of van Welsenaere and Froment, the latter maximum defines the critical conditions for runaway, i.e., it provides the maximum value of Λ that allows one to have an easily controlled temperature in the reactor for any given set of the remaining parameters. In Fig. 4.8, the critical point on curve 1 is found at $\Lambda_c = 0.7$.

For the sake of comparison, critical conditions according to Adler and Enig are reported on the same figure. These are given by the curve with $\Lambda = 1.16$, which is the curve where two (overlapping) inflection points first appear, and this shows that the Adler and Enig criterion is less conservative when compared to the criterion of van Welsenaere and Froment. Moreover, it can be observed that all curves

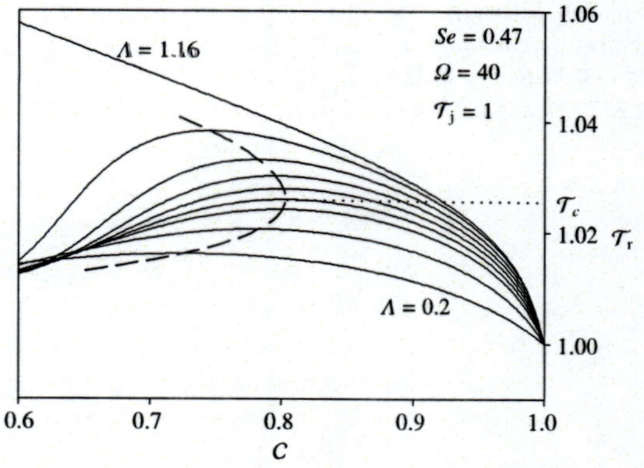

Fig. 4.8 Temperature–concentration profiles (*continuous lines*) at Λ values of (starting from the lowest curve) 0.2, 0.4, 0.6, 0.7, 0.8, 0.9, 1, 1.1, and 1.16 and locus of temperature maxima (*dashed line*). Critical conditions occur at $\Lambda_c = 0.7$ and $\mathcal{T}_c = 1.026$

$\mathcal{T}_r(\mathcal{C})$ reported in Fig. 4.8, and corresponding to safe operative conditions according to the Adler and Enig criterion, are obtained with the Semenov number $Se = 0.47$, which is well above the maximum critical value provided by the Semenov criterion, $Se_c = 0.377$, as given by (4.28). This difference mainly arises from the inclusion into the mathematical model of the terms accounting for consumption of reactant A.

The critical temperature $\mathcal{T}_c$ in Fig. 4.8 can be computed by observing that the concentration values corresponding to the temperature maxima $\mathcal{C}_{max}$ must satisfy the condition

$$\frac{d\mathcal{T}_r}{d\mathcal{C}} = 0, \tag{4.37}$$

which gives

$$\mathcal{C}_{max} = \frac{\Phi}{\Lambda} \exp\left[\Omega\left(\frac{1-\mathcal{T}_r}{\mathcal{T}_r}\right)\right](\mathcal{T}_r - \mathcal{T}_j). \tag{4.38}$$

The temperature $\mathcal{T}_c$ can be now calculated as the value that maximizes $\mathcal{C}_{max}$, i.e., by imposing the condition

$$\frac{d\mathcal{C}_{max}}{d\mathcal{T}_r} = 0, \tag{4.39}$$

which yields

$$\mathcal{T}_c = \frac{\Omega}{2}\left(1 - \sqrt{1 - \frac{4\mathcal{T}_j}{\Omega}}\right). \tag{4.40}$$

Interestingly, this is the same critical temperature predicted by the Semenov criterion (4.21). Again, a necessary condition for stability is $\mathcal{T}_{r0} < \mathcal{T}_c$, and, for the conditions of Fig. 4.8, it results in $\mathcal{T}_c = 1.026$.

On decreasing Se and keeping $\mathcal{T}_j = 1$, the locus of maxima moves to the right until it becomes tangent to the vertical axis $\mathcal{C} = 1$, as shown by curves 1, 2, and 3 in Fig. 4.9 obtained at $Se = 0.47$, 0.4, and 0.377, respectively. The tangency at $\mathcal{C} = 1$ occurs when the Semenov number is equal to the critical value predicted by the Semenov theory; indeed, by substituting $\mathcal{C}_{max} = 1$ and $\mathcal{T}_r = \mathcal{T}_c$ into (4.38), the same

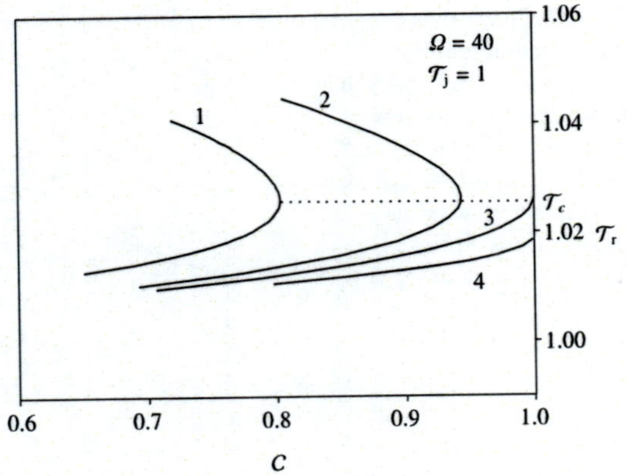

Fig. 4.9 Locus of temperature maxima at $Se = 0.47$ (curve *1*), 0.4 (*2*), 0.377 (*3*), and 0.35 (*4*)

relationship (4.19) is obtained. At any lower Se, the thermal stability of the system is always guaranteed, regardless of the value of Λ, and the temperature maxima are always less than $\mathcal{T}_c$, so that only the lower branch of the curves can be defined.

For the sake of brevity, the effects of $\mathcal{T}_{r0}$ and $\mathcal{T}_j$ on the runaway boundaries predicted by the different criteria are not discussed here and can be found in more details in the relevant literature (e.g., in [3, 14, 15]).

4.4.3 Sensitivity-based Runaway Criteria

A more recent class of criteria to identify runaway conditions is based on the evidence that, close to runaway boundaries, the system behavior becomes highly sensitive to even small changes of the model parameters. Within this class, the generalized criterion of Morbidelli and Varma [14] makes use of the absolute value of the normalized *objective* sensitivity s of the temperature maximum $\mathcal{T}_{r,max}$ to the generic model parameter θ, defined as

$$s(\mathcal{T}_{r,max}, \theta) = \left| \frac{\theta}{\mathcal{T}_{r,max}} \frac{\partial \mathcal{T}_{r,max}}{\partial \theta} \right|. \tag{4.41}$$

The criterion locates critical conditions where the values of all objective sensitivities to any model parameter θ reach a maximum. This is also a necessary condition for runaway, so that fast reactions that do not meet this requirement are not classified by Morbidelli and Varma as proper cases of thermal runaway.

Since an analytical expression for $s(\mathcal{T}_{r,max}, \theta)$ is usually difficult to obtain, it is more convenient to use the sensitivities of the reactor temperature $\mathcal{T}_r$ to the parameter θ as a function of time,

$$s(\mathcal{T}_r, \theta) = \left| \frac{\theta}{\mathcal{T}_r} \frac{\partial \mathcal{T}_r}{\partial \theta} \right|, \tag{4.42}$$

so that $s(\mathcal{T}_{r,max}, \theta)$ can be computed as the value of $s(\mathcal{T}_r, \theta)$ obtained for $\mathcal{T}_r = \mathcal{T}_{r,max}$.

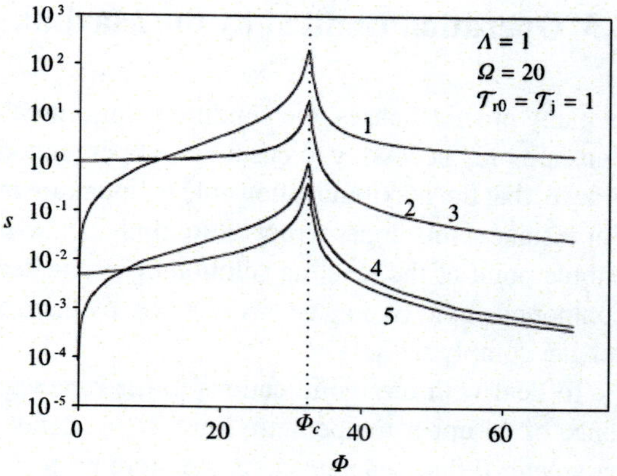

Fig. 4.10 Absolute sensitivities of temperature maximum with respect to $\mathcal{T}_j$ (curve *1*), Λ (*2*), $\mathcal{T}_{r0}$ (*3*), Ω (*4*), and Φ (*5*) as a function of Φ. Here curves 2 and 3 overlap

The computation of the sensitivities in (4.42) is carried out either by approximating the derivatives with finite differences or by the method already introduced in Sect. 3.6, i.e., by integrating the differential equations

$$\frac{\partial s}{\partial \tau} = \frac{\partial}{\partial \tau}\left|\frac{\theta}{\mathcal{T}_r}\frac{\partial \mathcal{T}_r}{\partial \theta}\right| \tag{4.43}$$

together with the reactor mass and energy balances ((4.11) and (4.12)) and the analogous time derivative of the sensitivities of $\mathcal{C}$, $s(\mathcal{C}, \theta)$.

In this way, the absolute normalized sensitivities $s(\mathcal{T}_{r,\max}, \theta)$ as a function of any model parameter, e.g., Φ, can be computed. This is shown in Fig. 4.10, where all sensitivities have a maximum at $\Phi_c = 32.45$, which consequently defines the critical limit for this parameter in order to guarantee an explosion-free operation.

Following the Morbidelli and Varma criterion, several other methods have been proposed in recent years in order to characterize the highly sensitive behavior of a batch reactor when it reaches the runaway boundaries. Among the most successful approaches, the evidence of a volume expansion in the phase space of the system has been widely exploited to characterize runaway conditions. For example, Strozzi and Zaldivar [9] defined the sensitivity as a function of the sum of the time-dependent Lyapunov exponents of the system and the runaway boundaries as the conditions that maximize or minimize this Lyapunov sensitivity. This has put the basis for the development of a new class of runaway criteria referred to as divergence-based approaches [5, 10, 18]. These methods usually identify runaway with the occurrence of a positive divergence of the vector field associated with the mathematical model of the reactor.

Finally, the novel stretching-based approach recently proposed by Adrover et al. [2] provides a formal bridge between the geometry- and the divergence-based runaway criteria. After introducing the so-called tangential, normal, and overall tangential stretching rates, runaway conditions are defined as the occurrence of a point along a system orbit at which the normalized tangential acceleration is positive and, moreover, the vector field experiences an overall stretching with respect to the initial value.

4.5 Operation Limited by the Maximum Allowable Temperature

In many practical cases, the conditions for criticality described in the previous sections are only necessary to ensure safe operation. Such conditions do not guarantee, indeed, that the maximum allowable temperature in the reactor, $\mathcal{T}_{ma}$, is not exceeded. For instance, this upper temperature limit can be imposed, in liquid systems, by the bubble point of the reacting solution or by the decomposition temperature of some compounds in it, or, in gaseous systems, by the maximum internal pressure the vessel can comply with.

To deal with the modifications to the runaway criteria determined by the existence of an upper temperature limit, let us denote by θ_{ma} the value of the generic parameter θ that generates a temperature profile with a maximum equal to $\mathcal{T}_{ma}$. In the same way, let us denote by θ_c the critical value of the parameter corresponding to the runaway boundaries obtained with any of the criteria discussed in the previous sections and generating a temperature profile with maximum value $\mathcal{T}_c$. Two possible cases can be distinguished, namely

$$\mathcal{T}_c < \mathcal{T}_{ma}, \tag{4.44}$$

$$\mathcal{T}_c > \mathcal{T}_{ma}. \tag{4.45}$$

When condition (4.44) is satisfied, the existence of a maximum allowable temperature does not affect the critical parameter value θ_c obtained with any of the classical runaway criteria. On the contrary, when condition (4.45) is satisfied, the actual limit value of the parameter to define safe operation must be replaced by θ_{ma}.

Regardless of which one of the previous conditions is met, when θ has a monotonically positive effect on the maximum reactor temperature $\mathcal{T}_{ma}$, as is the case of Λ, Ω, and $\mathcal{T}_{r0}$, the following scenarios can be encountered [7]:

1. $\theta < \theta_c$ and $\theta < \theta_{ma}$: safe.
2. $\theta > \theta_c$ and $\theta < \theta_{ma}$: sensitive.
3. $\theta < \theta_c$ and $\theta > \theta_{ma}$: unsafe.
4. $\theta > \theta_c$ and $\theta > \theta_{ma}$: unsafe.

If θ has a monotonically negative effect on $\mathcal{T}_{ma}$, as for $\mathcal{T}_j$ and Φ, all previous inequalities must be inverted.

The second condition (sensitive reaction) denotes a situation where runaway technically occurs, but the temperature peak does not exceed $\mathcal{T}_{ma}$, so that it can be still considered as a safe condition. This circumstance is, however, unfavorable in all cases when it is important to keep the reactor at nearly constant setup temperature during the entire reaction cycle.

The four scenarios were originally discussed by Fortuin et al. [7] with reference to the effect of variation of the dimensionless adiabatic temperature raise, but their validity can be extended to other model parameters as far as their effect on the maximum reactor temperature is monotonic.

4.6 Case Study: Runaway Boundaries

The phenol–formaldehyde reactive system introduced in Sect. 2.4 represents an interesting case study to highlight some aspects concerning the runaway boundaries in complex exothermic reaction networks occurring in the presence of a liquid phase. Since the set-point temperature for the production of substituted phenols usually lies in the range 60–90°C and water is used as solvent, under runaway conditions the boiling temperature of the liquid phase is easily reached: at such a point, evaporation of the solution takes place, and the internal reactor pressure starts to increase. If the reaction is not promptly stopped by means of thermal or chemical quenching, this can cause explosion of the reactor or the opening of relief valves with possible release of the reactor content into the working environment.

Because of the aforementioned circumstances, the loss of control of the phenol–formaldehyde reaction has been the cause of a number of severe incidents in chemical batch reactors during the last decades [12]. These incidents have caused many injuries and, in the worst case, even fatalities among the plant operators. Other severe consequences have been the evacuation of residents in the surrounding area due to chemical contamination and a protracted stop in the plant production.

For a safe operation, the runaway boundaries of the phenol–formaldehyde reaction must be determined. This is done here with reference to an isoperibolic batch reactor (while the temperature-controlled case is addressed in Sect. 5.8). As shown in Sect. 2.4, the complex kinetics of this system is described by 89 reactions involving 13 different chemical species. The model of the system consists of the already introduced mass (2.27) and energy (2.30) balances in the reactor. Given the system complexity, dimensionless variables are not introduced.

Safety boundaries have been determined according to three main approaches based on geometric features, parametric sensitivity, and maximum allowable temperature in the reactor, respectively.

As concerns the first approach, the Thomas and Bowes criterion has been adopted since it makes use of the second and third derivatives of the temperature-time profile, which are easy to obtain numerically. On the contrary, the Adler and Enig and the van Welsenaere and Froment criteria were discarded because of the difficulty in identifying a reference temperature–concentration profile, given the presence of many different reactants and intermediates.

For the sensitivity-based approach, the classical criterion of Morbidelli and Varma has been adopted. However, analytical computation of the sensitivities with respect to the model parameters (4.43) are not approachable, because of the high model complexity. Hence, the sensitivities have been estimated numerically by applying the following central approximation for the derivative:

$$s(\mathcal{T}_{\mathrm{r,max}}, \theta) = \frac{\theta}{\mathcal{T}_{\mathrm{r,max}}(\theta)} \left| \frac{\mathcal{T}_{\mathrm{r,max}}(\theta + h) - \mathcal{T}_{\mathrm{r,max}}(\theta - h)}{2h} \right|, \tag{4.46}$$

where h is the step used for incrementing θ (so that the global error of this approximation is $O(h^2)$).

Fig. 4.11 Safety boundaries for the phenol–formaldehyde reaction according to the runaway criteria of Morbidelli and Varma (*1*), Thomas and Bowes (2), and for an imposed maximum allowable reactor temperature $T_{r,ma} = 98°C$ (*3*)

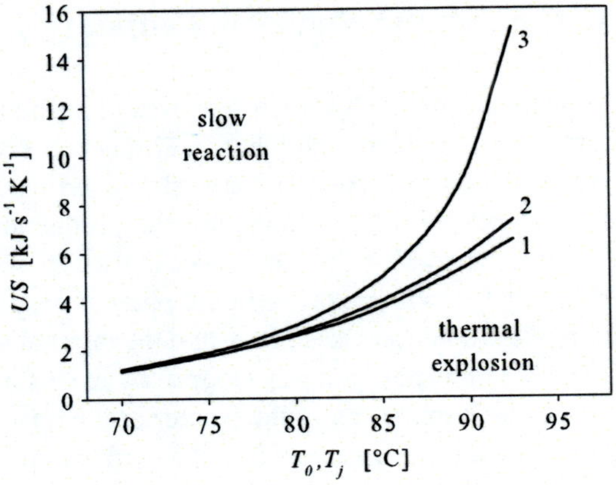

Finally, as concerns the third approach, the maximum allowable temperature in the reactor is determined by the boiling point of the reacting solution. This bubbling point is not easy to estimate theoretically since it refers to a nonideal solution containing many different species, and, moreover, it is likely to increase during the batch cycle, due to the increase of the mean molecular weight of the phenolic compounds. For the sake of simplicity, the boiling point of the reactive solution was supposed to be 98°C, which can be considered as an acceptable precautionary value when referred to the early stages of the batch cycle.

The simulations have been focused on the effect of the heat exchange equipment on reactor stability. To this goal, the minimum value of the group US that guarantees a thermally stable operation has been determined as a function of the initial reactor temperature T_{r0} (where $T_j = T_{r0}$ has been imposed). The critical US values are reported in Fig. 4.11 and, regardless of the particular criterion adopted, they increase—as expected—when T_{r0} increases.

Interestingly, the three safety criteria converge when the temperature is decreased below 75°C, while considerable differences in the definition of the safe operative conditions are observed in the higher temperature range (80–95°C). Here, the Morbidelli and Varma criterion is the less conservative one since it predicts stability with the lowest US critical values. The Thomas and Bowes criterion predicts, however, a critical US that is only slightly larger, so that the two results can be considered as nearly equivalent in practice. Above 80°C, both curves are largely overwhelmed by the US critical values imposed by the prevention of the maximum allowable temperature $T_{r,ma} = 98°C$ (curve 3). This means that, with reference to the particular chemical system considered here and under the investigated operative conditions, the safety boundaries for the phenol–formaldehyde reaction are determined by the need to prevent evaporation of the liquid, rather than by the onset of peculiar conditions indicated by the different runaway criteria available in the literature.

4.7 Conclusions

The discussion in the present chapter has dealt with the determination of safe operative conditions for exothermic batch reactors. To this aim, classical criteria have been summarized to define reactor runaway boundaries, and they eventually provide different results in terms of critical model parameters. For a deeper understanding of such discrepancies, it must be considered that, when abandoning the mere bifurcational problem addressed by the Semenov theory, the definition of the conditions at which runaway onsets is mostly a matter of convention. There is, indeed, no discontinuity in the reactor behavior when going from slow to fast reaction by gradually changing some of the reactor parameters. This does not much affect results in reactive systems with high B numbers where, given the very high sensitivity of the system, the transition from slow to fast reaction is always abrupt and, in practical cases, can still be treated as a discontinuity. Consequently, in this case almost all the approaches described here provide similar runaway boundaries.

However, several exothermic reactions are characterized by moderate or low values of the B number: here, the transition stages from safe to runaway conditions may cover a quite wide range of the parameter values, and the choice of the boundaries for the safe region is very discretional. Hence, not surprisingly, the main discrepancies among the different criteria are found at low B numbers [14, 15]. Moreover, in this case, runaway is a less dramatic phenomenon posing the problem to decide whether a bland explosion still represents a safety issue. In this case, an effective runaway criterion should be more properly determined on the basis of the actual ability of the system to comply with certain levels of temperature and pressure.

Moreover, in some practical cases, as highlighted by the case study, the observance of a maximum allowable temperature may even nullify the results deriving from the direct application of the runaway criteria. In detail, this happens when operation that is close, but not necessarily beyond, the runaway boundaries already produces a maximum temperature that, for some reason, the system cannot comply with. Hence, warnings are given about the necessity to include this aspect when investigating safe conditions at which exothermic reactions are to be carried out.

References

1. J. Adler and J.W. Enig. The critical conditions in thermal explosion theory with reactant consumption. *Combustion and Flame*, 8:97–103, 1964.
2. A. Adrover, F. Creta, M. Giona, and M. Valorani. Explosion limits and runaway criteria: a stretching-based approach. *Chemical Engineering Science*, 62:1171–1183, 2007.
3. V. Balakotaiah, D. Kodra, and D. Nguyen. Runaway limits for homogeneous and catalytic reactors. *Chemical Engineering Science*, 50:1149–1171, 1995.
4. J.A. Barton and P.A. Nolan. Incidents in the chemical industry due to thermal runaway chemical reactions. In *Proc. Conference and Exhibition on Techniques for Assessment of Chemical Reaction Hazards*, IBC, London, 1989.
5. J. Bosch, F. Strozzi, J.P. Zbilut, and J.M. Zaldívar. On-line runaway detection in isoperibolic batch and semibatch reactors using the divergence criterion. *Computers and Chemical Engineering*, 28:527–544, 2004.

6. J.C. Etchells. Prevention and control of exothermic runaway. In *Proc. Conference on Assessment and Control of Chemical Reaction Hazards*, IBC, London, 1993.
7. J.M.H. Fortuin, J.J. Heiszwolf, and C.S. Bildea. Design procedure for safe operations in agitated batch reactors. *AIChE Journal*, 47(3):920–928, 2001.
8. N.N. Semenov. Zur Theorie des Verbrennungsprozesses. *Zeitschrift für Physik*, 48:571–582, 1928.
9. F. Strozzi and J.M. Zaldívar. Thermal stability of batch chemical reactors by sensitivity calculation based on Lyapunov exponents. *Chemical Engineering Science*, 49:2681–2688, 1994.
10. F. Strozzi, J.M. Zaldívar, A.E. Kronberg, and K.R. Westerterp. On-line runaway detection in batch reactors using chaos theory techniques. *AIChE Journal*, 45(11):2429–2443, 1999.
11. P.H. Thomas and P.C. Bowes. Some aspects of the self-heating and ignition of solid cellulosic materials. *British Journal of Applied Physics*, 12:222–229, 1961.
12. U.S. Environmental Protection Agency. How to prevent runaway reactions. Document 550-F99-004, 1999.
13. R.J. van Welsenaere and G.F. Froment. Parametric sensitivity and runaway in fixed bed catalytic reactors. *Chemical Engineering Science*, 25:1503–1516, 1970.
14. A. Varma, M. Morbidelli, and H. Wu. *Parametric Sensitivity in Chemical Systems*. Cambridge University Press, Cambridge, 1999.
15. E. Velo, C.M. Bosch, and F. Recasens. Thermal safety of batch reactors and storage tanks. Development and validation of runaway boundaries. *Industrial & Engineering Chemistry Research*, 35(4):1288–1299, 1996.
16. K.R. Westerterp and E.J. Molga. Safety and runaway prevention in batch and semibatch reactors—A review. *Chemical Engineering Research and Design*, 84(A7):543–552, 2006.
17. H. Wu, M. Morbidelli, and A. Varma. An approximate criterion for reactor thermal runaway. *Chemical Engineering Science*, 53:3341–3344, 1998.
18. J.M. Zaldívar, J. Cano, M.A. Alós, J. Sempere, R. Nomen, D. Lister, G. Maschio, T. Obertopp, E.D. Gilles, J. Bosch, and F. Strozzi. A general criterion to define runaway limits in chemical reactors. *Journal of Loss Prevention in the Process Industries*, 16:187–200, 2003.

Chapter 5
Model-based Control

List of Principal Symbols

A	phenol (reduced model)
$\boldsymbol{a}(\boldsymbol{y})$	vector defined in (5.18)
$\boldsymbol{A}(\boldsymbol{y})$	matrix defined in (5.16)
a_{q}	scalar quantity defined in (5.22) [K s^{-1}]
$\boldsymbol{A}_{\mathrm{E}}(\boldsymbol{y})$	matrix defined in (5.28)
$\boldsymbol{A}_{\mathrm{M}}(\boldsymbol{y})$	matrix defined in (5.17)
$\boldsymbol{A}_{\mathrm{ME}}(\boldsymbol{y})$	matrix defined in (5.18)
$\boldsymbol{b}(\boldsymbol{y}, u)$	vector defined in (5.19)
$\boldsymbol{b}_{\mathrm{E}}(\boldsymbol{y}, u)$	vector defined in (5.19)
c	mass heat capacity [J kg^{-1} K^{-1}]
C	concentration [mol]
$\boldsymbol{C}$	matrix defined in (5.21)
$\boldsymbol{e}$	error vector between desired and measured output [K]
$\boldsymbol{f}, \boldsymbol{g}, h$	functions defining the nonlinear state space model (5.5)
g	controller gain
ΔH_{R}	molar enthalpy change of reaction [J mol^{-1}]
I	reaction intermediate (reduced model)
$\boldsymbol{I}_{m\times n}$	$m \times n$ identity matrix
k_{c}	rate constant
l	scalar gains of the observers
$\boldsymbol{L}$	matrix gains of the observers
$\boldsymbol{N}$	positive definite matrix defined in (5.43)
N_{C}	number of compounds involved in the reaction
N_{S}	number of time steps
N_{φ}	number of radial basis functions
$\boldsymbol{O}_{m\times n}$	$m \times n$ null matrix
P	desired product (reduced model)
$\boldsymbol{P}_{\mathrm{c}}$	positive definite matrix defined in (5.42)
S	heat transfer area [m^2]
t	time [s]

F. Caccavale et al., *Control and Monitoring of Chemical Batch Reactors*,
Advances in Industrial Control,
DOI 10.1007/978-0-85729-195-0_5,

T	temperature [K]
T_{in}	temperature of the fluid entering the jacket [K]
Δt	sampling time [s]
u	control input variable
U	overall heat transfer coefficient [$\mathrm{J\,m^{-2}\,K^{-1}\,s^{-1}}$]
$\boldsymbol{u}$	vector of control input variables
v	transformed input in (5.6)
V	volume [$\mathrm{m^3}$]
$\boldsymbol{x}$	vector of state variables
y	measured output variable
$\boldsymbol{y}$	vector of measured output variables
$\|\cdot\|$	Euclidean norm

Greek Symbols

α	parameters defined in (5.20)
β_{j}	parameter defined in (5.19)
γ	gain setting the parameter estimate update rate
$\boldsymbol{\varepsilon}$	error vector defined in (5.41)
θ	parameter $U\,S$
$\boldsymbol{\theta}_{\mathrm{q}}$	vector of unknown parameters defined in (5.32)
κ	centroid of an RBF
$\lambda(\cdot)$	eigenvalue of a matrix
$\boldsymbol{\xi}$	vector defined in (5.28)
ρ	density [$\mathrm{kg\,m^{-3}}$]
ς	interpolation error of the RBFI
υ	stoichiometric coefficient
ϕ_x, ϕ_v	nonlinear functions in (5.6)
$\boldsymbol{\varphi}$	vector radial basis functions defined in (5.32)
$\boldsymbol{\chi}_{\mathrm{c}}$	vector defined in (5.41)
$\boldsymbol{\psi}$	vector defined in (5.20)
$\boldsymbol{\psi}_{\mathrm{c}}$	vector defined in (5.48)
ω	width of an RBF

Subscripts and Superscripts

c	controller
D	derivative term (PID)
des	desired
E	energy balance
F	formaldehyde
I	integral term (PID)
j	jacket
M	mass balance
max	maximum value
min	minimum value
o	observer

P	proportional term (PID)
Ph	phenol
r	reactor
0	initial conditions
$\hat{}$	estimate
$\tilde{}$	estimation error

5.1 Control Strategies for Batch Reactors

The main goal of a control system for chemical batch reactors is that of imposing a given temperature–time profile inside the vessel. The effectiveness of the control action in tracking the desired temperature during the different reaction phases is usually a critical aspect in the chemical industry, since the temperature inside the reactor vessel strongly affects the reactor productivity and the quality of the final products [45]. Moreover, temperature control may become critical even for plant and operators safety since, in the presence of exothermic reactions, the heat generated by reaction can exceed the rate of heat exchanged with the cooling medium and this can cause reactor runaway.

Early approaches to control of chemical processes were mainly based on linear methods. Among them, the classical proportional-integral-derivative (PID) regulator [12] is still frequently adopted in the industrial practice since it is characterized by a simple structure and its tuning does not require knowledge of the mathematical model of the process. If the process is only mildly nonlinear or operates near a nominal steady-state condition, linear PID controllers can provide satisfactory performance. However, industrial chemical processes are often characterized by a strong nonlinear behavior and/or operate within a wide range of conditions. In these cases, PID controllers must be tuned very conservatively in order to guarantee a stable behavior over the entire range of operation; of course, such a conservative tuning usually causes degradation of the overall control system performance [31].

The control performance can be usually improved by incorporating the mathematical model of the process into the control law. Starting from the last decades of the 20th century, some promising model-based control strategies, e.g., the Model Predictive Control (MPC) [57], based on linear or linearized models, have been proposed. MPC is an optimal control method, which computes the control inputs by minimizing an objective function. This function is defined in terms of both present and predicted system variables and is evaluated by using an explicit model to predict future process outputs. Since MPC is based on the use of linear or linearized models, it may suffer from the same drawbacks of the PID regulator unless more advanced nonlinear MPC techniques are adopted [46]. Hence, in the last two decades, nonlinear model-based control strategies have been largely investigated. These techniques use a nonlinear model of the process to compute the control output.

The most general approach to model-based nonlinear control is the so-called Feedback Linearization (FL) [35]. In fact, FL control approaches use the model of the plant to achieve a global linearization of the closed-loop systems, so as well-established linear controllers can be adopted for the globally linearized model. In

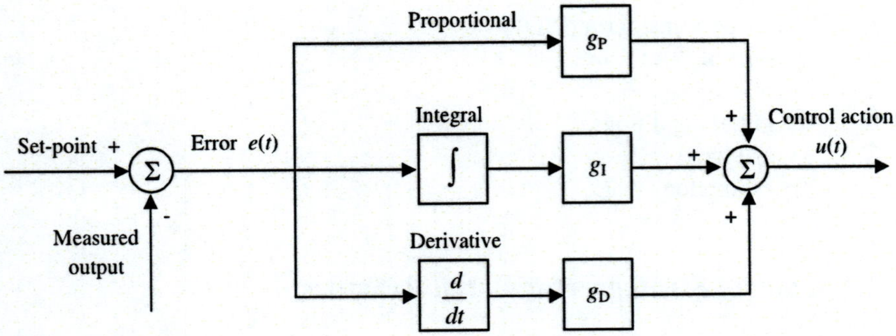

Fig. 5.1 PID regulator

its basic formulation, FL control needs an accurate mathematical model and the measurement of the whole state of the controlled plant.

In order to tackle the problem of uncertainties in the available model, nonlinear robust and adaptive strategies have been developed, while, in the absence of full state measurements, output-feedback control schemes can be adopted, where the unmeasurable state variables can be estimated by resorting to state observers. The development of model-based nonlinear strategies has been fostered by the development of efficient experimental identification methods for nonlinear models and by significantly improved capabilities of computer-control hardware and software.

In the following, an overview of the above discussed linear and nonlinear approaches to temperature control of batch reactors is provided.

5.2 PID Regulator

The PID regulator is the most widely used feedback controller in industrial settings. The output of a PID controller is given by the composition of three different correcting terms (Fig. 5.1): a proportional term, an integral term, and a derivative term, i.e.,

$$u(t) = g_P e(t) + g_I \int_{t_0}^{t} e(\zeta)\, d\zeta + g_D \dot{e}(t), \tag{5.1}$$

where u is the controller output (control action), e is the error between the desired output (set-point) and the measured output, g_P, g_I, and g_D are positive gains, and ζ is the integration variable, which represents the time elapsed since the initial time t_0.

The gains of the controller must be chosen in order to stabilize robustly the closed-loop system and, at the same time, to adjust the transient response (e.g., in terms of overshoot, rising time, and settling time). Often, these are conflicting requirements.

Since actuators are subject to saturation, the phenomenon of integral *windup* must be properly tackled; namely, if the controller outputs a command beyond the actuator saturation limits, the tracking error grows and the output of the integrator becomes very large; hence, unfavorable behaviors of the closed-loop system may arise (e.g., large overshoot and/or long settling time). Several anti-windup mechanism have been proposed to counteract this phenomenon [12].

The success of PID controller in industry is partly due to the existence of automatic tuning methods, requiring a very approximate (or null) knowledge of the plant. Early tuning methods, due to Ziegler and Nichols, have evolved over the years, and nowadays many other effective approaches are available [12]. More recently, many attempts have been made at modifying the basic PID structure in such a way to take into account the nonlinearities of the process under control. In [17, 33, 59], self-tuning PID controllers have been proposed, based on optimization criteria or adaptive techniques, while in [3] the PID design parameters are found via a genetic algorithm.

5.3 Model Predictive Control

The Model Predictive Control has been characterized by a wide success in industrial applications. It requires a model of the process (and, eventually, an estimate of the disturbances), the measurement on a given time horizon of both the input (i.e., the control action $\boldsymbol{u}$) and the output (i.e., the controlled variable $\boldsymbol{y}$) of the controlled process, the desired output ($\boldsymbol{y}_{\text{des}}$), and a prediction of the process input and output on the same time horizon.

MPC is traditionally formulated directly in the discrete-time domain, i.e., for sampled data systems. In detail, it is assumed that the system's input is computed (and the system's output is measured) only at discrete time steps $t_k = k\Delta t$, where k is an integer variable, and Δt is the sampling time. Hereafter, for the sake of compactness, the discrete-time variable is denoted simply by the integer k.

The basic idea of MPC is to compute, at each time step, k, a prediction of the control action values in N_S time steps, $\boldsymbol{u}(k+i \mid k)$ $(i = 1, \dots, N_S)$, where the notation $i|j$ stands for the value computed at step i on the basis of the information available up to step j. Then, by using the process model and the predicted control action, an estimate of the process output in the same N_S time steps $\boldsymbol{y}(k+i \mid k)$ $(i = 1, \dots, N_S)$ is computed. The predicted value at time $k+1$ is applied to the process.

The prediction of the control input is computed via an optimization method that minimizes a suitably defined objective function, usually composed by two terms: the first one is related to the deviation of the predicted output from the reference trajectory (i.e., the tracking error), while the second term takes into account control input changes. Hence, the optimization problem has the form

$$\min_{\substack{\boldsymbol{u}(k+i|k)\\ i\in[1,\dots,N_S]}} \sum_{i=1}^{N_S} \left\| \boldsymbol{y}_{\text{des}}(k+i) - \boldsymbol{y}(k+i \mid k) \right\|^2 + \sum_{i=1}^{N_S} \left\| \Delta\boldsymbol{u}(k+i \mid k) \right\|^2, \tag{5.2}$$

where $\|\cdot\|$ denotes the usual Euclidean norm, and

$$\Delta\boldsymbol{u}(k+i \mid k) = \boldsymbol{u}(k+i \mid k) - \boldsymbol{u}(k+i-1 \mid k). \tag{5.3}$$

In this framework, constraints on the decision variables can be directly taken into account, i.e., the optimization problem is (5.2) subject to

$$\begin{aligned} \boldsymbol{u}_{\min} &\le \boldsymbol{u}(k+i \mid k) \le \boldsymbol{u}_{\max}, \\ \Delta\boldsymbol{u}_{\min} &\le \Delta\boldsymbol{u}(k+i \mid k) \le \Delta\boldsymbol{u}_{\max}, \\ \boldsymbol{y}_{\min} &\le \boldsymbol{y}(k+i \mid k) \le \boldsymbol{y}_{\max}. \end{aligned} \tag{5.4}$$

It can be easily argued that the choice of the process model is crucial to determine the nature and the complexity of the optimization problem. Several models have been proposed in the literature, ranging from simple state-space linear models to complex nonlinear mappings. In the case where a linear model is adopted, the objective function to be minimized is quadratic in the input and output variables; thus, the optimization problem (5.2), (5.4) admits analytical solutions. On the other hand, when nonlinear models are used, the optimization problem is not trivial, and thus, in general, only suboptimal solutions can be found; moreover, the analysis of the closed-loop main properties (e.g., stability and robustness) becomes more challenging.

Early applications of MPC took place in the 1970s, mainly in industrial contexts, but only later MPC became a research topic. One of the first solid theoretic formulations of MPC is due to Richalet et al. [53], who proposed the so-called Model Predictive Heuristic Control (MPHC). MPHC uses a linear model, based on the impulse response and, in the presence of constraints, computes the process input via a heuristic iterative algorithm. In [23], the Dynamic Matrix Control (DMC) was introduced, which had a wide success in chemical process control; both impulse and step models are used in DMC, while the process is described via a matrix of constant coefficients. In later formulations of DMC, constraints have been included in the optimization problem. Starting from the late 1980s, MPC algorithms using state-space models have been developed [38, 43]. In parallel, Clarke et al. used transfer functions to formulate the so-called Generalized Predictive Control (GPC) [19–21] that turned out to be very popular in chemical process control. In the last two decades, a number of nonlinear MPC techniques has been developed [34, 46, 57].

As concerns the application of MPC to batch reactors, a number of works have been recently proposed [39, 48, 49]. A major challenge in designing MPC for batch processes is to find a model valid in any operating condition. To this purpose, a number of different solutions have been proposed, including, among the most interesting ones, adaptive techniques [32] and the use of multiple-model approaches [36]. In [55], a set of linearized models is adopted to design a nonlinear MPC for an industrial polypropylene semi-batch reactor process; then, a quadratic programming problem is applied to each local linearization of the nonlinear process to determine the actual control action.

The interested reader is referred to [1], where a wide review of MPC can be found with special emphasis on applications to chemical and biological processes.

5.4 Feedback Linearization

In order to briefly introduce the basic principle of the feedback linearizing control, consider the following Single Input Single Output (SISO) nonlinear model:

$$\begin{cases} \dot{\boldsymbol{x}} = \boldsymbol{f}(\boldsymbol{x}) + \boldsymbol{g}(\boldsymbol{x})u, \\ y = h(\boldsymbol{x}), \end{cases} \tag{5.5}$$

where, as usual, $\boldsymbol{x}$, u, and y denote the vector of the state variables, the plant input, and the plant output, respectively. The problem of the feedback linearizing control is to find a state feedback control of the form

$$u = \phi_x(\boldsymbol{x}) + \phi_v(\boldsymbol{x})v \tag{5.6}$$

such that the resulting closed-loop system is characterized by a linear dynamics [35]. In (5.6), v plays the role of new control input, while ϕ_x and ϕ_v are suitable nonlinear functions. Input–output feedback linearization is aimed at obtaining a linear map between the new control input v and the actual plant output y, whereas the state-space feedback linearization [31, 35], instead, is aimed at obtaining a linear map between the new control input v and the state $\boldsymbol{x}$. In both cases, the new input v can be designed by resorting to well-established linear control techniques.

Since, in process control, input–output linearization techniques are usually preferred to state-space approaches, mostly due to the higher complexity of the latter, in the following, only input–output feedback linearization basic concepts are briefly reviewed.

5.4.1 Input–Output Linearization

The time derivative of the output of system (5.5) can be written as

$$\dot{y} = \frac{\partial h}{\partial \boldsymbol{x}}\big[\boldsymbol{f}(\boldsymbol{x}) + \boldsymbol{g}(\boldsymbol{x})u\big] = L_f h + (L_g h)u, \tag{5.7}$$

where $L_f h$ is the Lie derivative of h with respect to $\boldsymbol{f}$ and is defined as

$$L_f h = \frac{\partial h}{\partial \boldsymbol{x}} \boldsymbol{f}(\boldsymbol{x}). \tag{5.8}$$

The same definition applies for $L_g h$, where the function g plays the same role as $\boldsymbol{f}$ in (5.8). Of course, higher-order Lie derivatives can be defined in a recursive way, e.g.,

$$L_f^2 h = \frac{\partial L_f h}{\partial x} \boldsymbol{f}(\boldsymbol{x}), \qquad L_g L_f h = \frac{\partial L_f h}{\partial x} \boldsymbol{g}(\boldsymbol{x}).$$

If the relative order μ of y with respect to u is 1,[1] then $L_g h \neq 0$, and the procedure stops here. Otherwise, the procedure is iterated until $L_g L_f^{\mu-1} h \neq 0$, i.e., until the derivative of order μ is computed:

$$y^{(\mu)} = L_f^{\mu} h + \big(L_g L_f^{\mu-1} h\big) u. \tag{5.9}$$

Therefore, if the control input u is chosen as

$$u = \frac{v - L_f^{\mu} h}{L_g L_f^{\mu-1} h}, \tag{5.10}$$

the new input–output map, $v - y$, is reduced to a chain of μ integrators

$$y^{(\mu)} = v, \tag{5.11}$$

i.e., to a linear system of order μ. For this system, a feedback controller can be designed by resorting to linear control techniques.

Several process control design methods, such as the Generic Model Control (GMC) [41], the Globally Linearizing Control (GLC) [37], the Internal Decoupling Control (IDC) [7], the reference system synthesis [8], and the Nonlinear Internal Model Control (NIMC) [29], are based on input–output linearization.

However, feedback linearizing control requires the knowledge of an accurate model of the process. Hence, in the presence of parametric model uncertainties, adaptive or robust control strategies have been proposed [4, 10, 18, 30]; in [47], model uncertainties are tackled by adopting an Artificial Neural Network (ANN) in conjunction with different linearizing control strategies.

5.4.2 Generic Model Control

The Generic Model Control (GMC) is a model-based control strategy developed by Lee and Sullivan in 1988 [41]. It can be shown that GMC is an input–output linearization technique for processes with unitary relative order [31].

The basic idea of GMC is that of imposing a desired closed-loop dynamics of the form

$$\dot{y}(t) = g_{\mathrm{P}}\big(y_{\mathrm{des}}(t) - y(t)\big) + g_{\mathrm{I}} \int_0^t \big(y_{\mathrm{des}}(\zeta) - y(\zeta)\big)\, d\zeta, \tag{5.12}$$

where g_{P}, g_{I} are positive gains, and y_{des} is the desired output of the controlled process. By comparing equations (5.7) and (5.12), the following equality can be imposed:

$$L_f h + (L_g h) u = g_{\mathrm{P}}(y_{\mathrm{des}} - y) + g_{\mathrm{I}} \int_0^t \big(y_{\mathrm{des}}(\zeta) - y(\zeta)\big)\, d\zeta, \tag{5.13}$$

[1] The relative order of y with respect to u is the smallest integer μ such that $L_g L_f^{\mu-1} h \neq 0$.

which leads to the control law

$$u(t) = \frac{g_{\mathrm{P}}(y_{\mathrm{des}}(t) - y(t)) + g_{\mathrm{I}} \int_0^t (y_{\mathrm{des}}(\zeta) - y(\zeta))\, d\zeta - L_f h}{L_g h}. \tag{5.14}$$

Since the dynamics of a batch reactor is characterized by a unitary relative order, the GMC law can be adopted [6, 14, 22, 40, 42, 65]. In order to cope with model uncertainties, adaptive GMC approaches have been developed [56, 60, 62]; in [27] some unknown quantities—namely, the effect of the heat released by the reaction and the heat transfer coefficient—are estimated by adopting the nonlinear adaptive observer proposed in [24]; in [63], an ANN-based GMC approach is presented for semi-batch processes with relative order higher than one.

5.5 State-Space Model for Control Design

Most of the above reviewed approaches, including those based on the mathematical model of the system, are developed and tested for simplified reaction networks and/or for particular reactive systems. Nonetheless, it is worthwhile to focus on more general reaction schemes and on the rigorous analysis of the main structural properties (i.e., stability and robustness) of the closed-loop system.

In the following, the model-based controller–observer adaptive scheme in [15] is presented. Namely, an observer is designed to estimate the effect of the heat released by the reaction on the reactor temperature dynamics; then, this estimate is used by a cascade temperature control scheme, based on the closure of two temperature feedback loops, where the output of the reactor temperature controller becomes the setpoint of the cooling jacket temperature controller. Model-free variants of this control scheme are developed as well. The convergence of the overall controller–observer scheme, in terms of observer estimation errors and controller tracking errors, is proven via a Lyapunov-like argument. Noticeably, the scheme is developed for the general class of irreversible nonchain reactions presented in Sect. 2.5.

In order to design the controller, the model developed in Sect. 2.5 is conveniently reformulated in a state-space form. The following assumptions on the reactive process and on the reactor are used to derive the state-space form of the model.

Assumption 5.1 The reactions involved in the process are exothermic and characterized by first-order kinetics; therefore, the mass balances of the N_{C} compounds involved in the reaction have the form (2.27).

Assumption 5.2 Both the reactor vessel and the jacket are considered perfectly mixed; therefore, the energy balances in the reactor and in the jacket can be written as in (2.30) and (2.31), respectively.

Assumption 5.3 The overall coefficient U of heat exchange between the reactor and the cooling jacket is assumed to be poorly known, as expected in practical applications.

Assumption 5.4 It is assumed that only the temperature measurements in the reactor and in the jacket are available, i.e., neither the concentrations nor the heat released by the reaction can be measured in real-time. As discussed in Sect. 2.7, the temperature can be usually controlled via the heat exchange between reactor and a heating/cooling fluid that circulates in a jacket surrounding the vessel or in a coil inside the vessel. In the following, it is assumed that the control system uses the inlet temperature of the fluid in the jacket as manipulated variable; nevertheless, the control approaches developed in this chapter can be easily extended to different control strategies.

Let us define the $(N_C + 2) \times 1$ state vector

$$x = \begin{bmatrix} x_M \\ x_E \end{bmatrix}, \qquad x_M = \begin{bmatrix} C_1 \\ C_2 \\ \vdots \\ C_{N_C} \end{bmatrix} = \begin{bmatrix} x_1 \\ x_2 \\ \vdots \\ x_{N_C} \end{bmatrix}, \qquad x_E = \begin{bmatrix} T_r \\ T_j \end{bmatrix} = \begin{bmatrix} x_{N_C+1} \\ x_{N_C+2} \end{bmatrix},$$

the control input

$$u = T_{in},$$

the output vector of measurable variables (see Assumption 5.4)

$$y = \begin{bmatrix} T_r \\ T_j \end{bmatrix} = \begin{bmatrix} y_1 \\ y_2 \end{bmatrix} = x_E,$$

and the parameter (assumed to be poorly known, as stated by Assumption 5.3)

$$\theta = US.$$

Then, (2.27), (2.30), and (2.31) can be rewritten in the following state-space form:

$$\begin{cases} \dot{x} = A(y)x + b(y, u) + C^T \psi(y)\theta, \\ y = Cx, \end{cases} \tag{5.15}$$

where the matrix A is given by

$$A = \begin{bmatrix} A_M(y) & O_{N_C \times 2} \\ A_{M,E}(y) & O_{2 \times 2} \end{bmatrix}, \tag{5.16}$$

$O_{m \times n}$ denotes the $m \times n$ null matrix, and

$$A_M(y) = \begin{bmatrix} -k_{c1}(T_r) & 0 & \dots & 0 \\ \upsilon_{1,2} k_{c1,2}(T_r) & -k_{c2}(T_r) & \dots & 0 \\ \vdots & \vdots & \vdots & \vdots \\ \upsilon_{1,p} k_{c1,N_C}(T_r) & \upsilon_{2,N_C} k_{c2,N_C}(T_r) & \dots & -k_{cN_C}(T_r) \end{bmatrix}, \tag{5.17}$$

$$A_{\mathrm{M,E}}(y) = \begin{bmatrix} a_1(T_\mathrm{r}) & \dots & a_{N_\mathrm{C}}(T_\mathrm{r}) \\ 0 & \dots & 0 \end{bmatrix} = \begin{bmatrix} \boldsymbol{a}^\mathrm{T}(y) \\ \boldsymbol{0}^\mathrm{T}_{N_\mathrm{C}\times 1} \end{bmatrix}, \tag{5.18}$$

$$a_i(y) = \sum_{h=i+1}^{N_\mathrm{C}+1} \alpha_{i,h} k_{\mathrm{c}i,h}(T_\mathrm{r}), \qquad \alpha_{i,h} = \frac{(-\Delta H_{\mathrm{R}i,h})}{\rho_\mathrm{r} c_\mathrm{r}}, \qquad i = 1, \dots, N_\mathrm{C}.$$

The matrix A_M is characterized by the following properties:

- each lumped overall rate constant of the reactions of disappearance, $k_{\mathrm{c}i}(T_\mathrm{r})$, defined in (2.29), is strictly positive (for all T_r) if compound A_i is involved at least in one reaction
- both $k_{\mathrm{c}i,h}$ and $k_{\mathrm{c}i}$ are upper and lower bounded as in (2.32) and (2.33), respectively; and
- the off-diagonal terms in A_M are nonnegative (for all T_r) and are null if the corresponding reaction, $\mathrm{A}_i \to \mathrm{A}_h$, does not take place (i.e., $\upsilon_{i,h} = 0$).

The vector $\boldsymbol{b}$ in (5.15) is defined by

$$\boldsymbol{b}(y,u) = \begin{bmatrix} \boldsymbol{0}_{N_\mathrm{C}\times 1} \\ \boldsymbol{b}_\mathrm{E}(y,u) \end{bmatrix}, \quad \boldsymbol{b}_\mathrm{E}(y,u) = \begin{bmatrix} 0 \\ \beta_\mathrm{j}(u - T_\mathrm{j}) \end{bmatrix}, \quad \beta_\mathrm{j} = \frac{F_\mathrm{V}}{V_\mathrm{j}}, \tag{5.19}$$

the vector $\boldsymbol{\psi}$ is given by

$$\boldsymbol{\psi}(y) = \begin{bmatrix} -\alpha_\mathrm{r}(T_\mathrm{r} - T_\mathrm{j}) \\ \alpha_\mathrm{j}(T_\mathrm{r} - T_\mathrm{j}) \end{bmatrix}, \quad \alpha_* = \frac{1}{V_* \rho_* c_*}, \quad * = \{\mathrm{r}, \mathrm{j}\}, \tag{5.20}$$

and the output matrix is given by

$$\boldsymbol{C} = \begin{bmatrix} \boldsymbol{O}_{2\times N_\mathrm{C}} & \boldsymbol{I}_{2\times 2} \end{bmatrix}, \tag{5.21}$$

where $\boldsymbol{I}_{n\times n}$ denotes the $n \times n$ identity matrix.

5.6 Estimation of the Heat Released by Reaction

From (5.15) it can be argued that the heat released by the reaction affects the dynamics of the reactor temperature via the term

$$a_\mathrm{q}(\boldsymbol{x}_\mathrm{M}, y) = \sum_{i=1}^{N_\mathrm{C}} \sum_{h=i+1}^{N_\mathrm{C}+1} \alpha_{i,h} k_{\mathrm{c}i,h}(T_\mathrm{r}) x_i = \sum_{i=1}^{N_\mathrm{C}} a_i(T_\mathrm{r}) x_i = \boldsymbol{a}^\mathrm{T}(y) \boldsymbol{x}_\mathrm{M}. \tag{5.22}$$

In turn, this term depends on the reactant concentrations, which are usually measurable at very low sampling rates, not suitable for real-time control. Hence, the design of a model-based control law for the reactor temperature should require an estimate of this term.

More in general, the use of observers for state estimation in batch reactors has been extensively investigated. In [61], an Extended Kalman Filter (EKF) is adopted to estimate the concentrations of reactants in an industrial semibatch reaction, assessing several critical issues related to the adoption of the EKF in industrial settings against open-loop models, caused by partial lack of observability and poor measurements quality, especially during the initial and final stages of the batch. More recent results [9], derived for a class of nonlinear differential-algebraic systems, showed more encouraging results, although these are obtained for a simulated batch reactor and a simple real mixing process. On the other hand, the heat released by the reaction can be estimated by adopting the approach known as *calorimetric method* [13, 54, 58], in which the energy balance is used together with measured values of temperature and its time derivative. In order to avoid numerical differentiation of the temperature measurements, an observer can be used to estimate both the effect of the heat released by the reaction and the heat-transfer coefficient [18, 27]. In [18], a nonlinear adaptive control strategy is adopted, based on an EKF to achieve online estimation of the time varying parameters involved in the control law, while in [27] the effect of the heat released by the reaction is estimated as an unknown parameter.

In the following, three different approaches to estimation of the effect of the heat released by the reaction on the system dynamics are presented: the first two are based on the results in [15, 51], while the third is one of the most interesting model-free approaches in the literature. In detail:

- The first approach adopts a nonlinear model-based adaptive observer to estimate the reactant concentrations (i.e., the state variables $x_1, \dots, x_{N_C}$), while the heat transfer coefficient is estimated via an adaptive update law. Then, the term a_q is reconstructed from the estimated concentrations.
- The second approach is based on a model-free estimator; the estimation is based on the adoption of a universal interpolator, i.e., a Radial Basis Function Interpolator (RBFI). Hence, differently from the previous approach, knowledge of the reaction kinetics is not required.
- The model-free approach proposed in [27], in which both the heat transfer coefficient and the thermal power are estimated as unknown parameters. As for the previous one, this approach does not need either the knowledge of the reaction kinetics or the estimation of the concentrations.

5.6.1 Model-Based Nonlinear Observer

The observer has the form

$$\begin{cases} \dot{\widehat{\boldsymbol{x}}} = \boldsymbol{A}(y)\widehat{\boldsymbol{x}} + \boldsymbol{b}(y,u) + \boldsymbol{L}\widetilde{\boldsymbol{y}} + \boldsymbol{C}^{\mathrm{T}}\boldsymbol{\psi}(y)\widehat{\theta}_{\mathrm{o}}, \\ \widehat{\boldsymbol{y}} = \boldsymbol{C}\widehat{\boldsymbol{x}}, \end{cases} \tag{5.23}$$

where $\widehat{\boldsymbol{x}}$ denotes the state estimate, while $\widehat{\boldsymbol{y}}$ and $\widetilde{\boldsymbol{y}} = \boldsymbol{y} - \widehat{\boldsymbol{y}}$ denote the output estimates and output estimation errors, respectively; $\boldsymbol{L}$ is an $(N_C + 2) \times 2$ matrix of positive gains

$$L = \begin{bmatrix} L_M \\ L_E \end{bmatrix}, \quad L_M = \begin{bmatrix} l_1 & 0 \\ l_2 & 0 \\ \vdots & \vdots \\ l_{N_C} & 0 \end{bmatrix}, \quad L_E = \begin{bmatrix} l_r & 0 \\ 0 & l_j \end{bmatrix},$$

and the estimate $\widehat{\theta}_o$ of θ is computed via the update law

$$\dot{\widehat{\theta}}_o = \gamma_o^{-1} \boldsymbol{\psi}^T(\boldsymbol{y}) \widetilde{\boldsymbol{y}}, \tag{5.24}$$

where γ_o is a positive gain setting the update rate.

Therefore, an estimate of a_q can be easily computed via (5.22) from the estimates of the reactants concentrations

$$\widehat{a}_q(\widehat{\boldsymbol{x}}_M, \boldsymbol{y}) = \sum_{i=1}^{N_C} \sum_{h=i+1}^{N_C+1} \alpha_{i,h} k_{ci,h}(T_r) \widehat{x}_i = \sum_{i=1}^{N_C} a_i(T_r) \widehat{x}_i = \boldsymbol{a}^T(\boldsymbol{y}) \widehat{\boldsymbol{x}}_M. \tag{5.25}$$

The convergence properties of both the state estimation error $\widetilde{\boldsymbol{x}} = \boldsymbol{x} - \widehat{\boldsymbol{x}}$ and the parameter estimation error $\widetilde{\theta}_o = \theta - \widehat{\theta}_o$ are stated in the following theorem.

Theorem 5.1 *If the rate constants are bounded as in* (2.32) *and* (2.33), *then, there exists a set of observer gains such that the state estimation error* $\widetilde{\boldsymbol{x}}$ *is globally uniformly convergent to* $\mathbf{0}$ *as* $t \to \infty$ *and the parameter estimation error* $\widetilde{\theta}_o$ *is bounded for every* t.

The proof is based on a Lyapunov-like argument and is reported in Appendix A.1.

Remark 5.1 As usual, in direct adaptive estimation and/or control schemes, the convergence to zero of the parameter estimation error $\widetilde{\theta}_o$ is not guaranteed, unless the *persistency of excitation* condition is fulfilled [5, 35]. In detail, if there exist three scalars $\lambda_1 > 0$, $\lambda_2 > 0$, and $\tau > 0$ such that

$$\lambda_1 \le \int_t^{t+\tau} \boldsymbol{\psi}^T\big(\boldsymbol{y}(\zeta)\big)\boldsymbol{\psi}\big(\boldsymbol{y}(\zeta)\big)\, d\zeta \le \lambda_2 \quad \forall t \ge 0, \tag{5.26}$$

then, both the state estimation error $\widetilde{\boldsymbol{x}}$ and the parameter estimation error $\widetilde{\theta}_o$ are globally exponentially convergent to zero.

The above remark is of the utmost importance for evaluating the potential of the proposed observer in a real setup. In fact, exponential stability would ensure robustness of the state estimation against bounded and/or vanishing model uncertainties and disturbances [35], due to inaccurate and/or incomplete knowledge of reaction kinetics and to usual simplifying assumptions adopted for the model derivation (e.g., perfect mixing).

Remark 5.2 In the case of perfect knowledge of θ, the observer takes the form (5.23), where the estimate $\widehat{\theta}_\mathrm{o}$ is replaced by the true value of the coefficient.

5.6.2 Model-Free Approaches

When an accurate model of the reaction kinetics is not available (e.g., due to the lack of reliable data for identification), the previously developed approach may be ineffective and model-free strategies for the estimation of the effect of the heat released by the reaction, a_q, must be adopted. In detail, the approach in [27] can be considered, where a_q is estimated, together with the heat transfer coefficient, via a suitably designed nonlinear observer [24]. Other model-free approaches can be adopted, e.g., those based on the adoption of universal interpolators (neural networks, polynomials) for the direct online estimation of the heat [16] and purely neural approaches [11]. Approaches based on the combination of neural and model-based paradigms [2] or on tendency models [25] can be considered as well.

In the following, two model-free approaches based on adaptive observer are presented: the first one is based on the results in [51], and the second one is the well-established observer proposed by [24] and applied to batch reactors in [27].

5.6.2.1 Approach Based on Universal Interpolators

In order to present this observer, the state space equation referred to the vector $\boldsymbol{x}_\mathrm{E}$ can be conveniently rewritten as

$$\begin{cases} \dot{\boldsymbol{x}}_\mathrm{E} = \boldsymbol{A}_\mathrm{E}(\theta)\boldsymbol{x}_\mathrm{E} + \boldsymbol{\xi}(\boldsymbol{x}_\mathrm{M}, \boldsymbol{x}_\mathrm{E}) + \boldsymbol{b}_\mathrm{E}(\boldsymbol{y}, u), \\ \boldsymbol{y} = \boldsymbol{x}_\mathrm{E}, \end{cases} \tag{5.27}$$

where $\boldsymbol{b}_\mathrm{E}$ has been defined in (5.19), and

$$\boldsymbol{A}_\mathrm{E}(\theta) = \begin{bmatrix} -\alpha_\mathrm{r}\theta & \alpha_\mathrm{r}\theta \\ \alpha_\mathrm{j}\theta & -\alpha_\mathrm{j}\theta \end{bmatrix}, \qquad \boldsymbol{\xi}(\boldsymbol{x}_\mathrm{M}, \boldsymbol{x}_\mathrm{E}) = \begin{bmatrix} a_\mathrm{q}(\boldsymbol{x}_\mathrm{M}, \boldsymbol{x}_\mathrm{E}) \\ 0 \end{bmatrix}. \tag{5.28}$$

The following observer can be adopted:

$$\begin{cases} \dot{\widehat{\boldsymbol{x}}}_\mathrm{E} = \boldsymbol{A}_\mathrm{E}(\widehat{\theta}_\mathrm{o})\widehat{\boldsymbol{x}}_\mathrm{E} + \widehat{\boldsymbol{\xi}}(\boldsymbol{\theta}_\mathrm{q}, \boldsymbol{y}) + \boldsymbol{b}_\mathrm{E}(\boldsymbol{y}, u) + \boldsymbol{L}_\mathrm{E}\widetilde{\boldsymbol{y}}, \\ \widehat{\boldsymbol{y}} = \widehat{\boldsymbol{x}}_\mathrm{E}, \end{cases} \tag{5.29}$$

where

$$\boldsymbol{A}_\mathrm{E}(\widehat{\theta}_\mathrm{o}) = \begin{bmatrix} -\alpha_\mathrm{r}\widehat{\theta}_\mathrm{o} & \alpha_\mathrm{r}\widehat{\theta}_\mathrm{o} \\ \alpha_\mathrm{j}\widehat{\theta}_\mathrm{o} & -\alpha_\mathrm{j}\widehat{\theta}_\mathrm{o} \end{bmatrix}, \quad \boldsymbol{L}_\mathrm{E} = \begin{bmatrix} l_\mathrm{r} & 0 \\ 0 & l_\mathrm{j} \end{bmatrix}, \quad \widehat{\boldsymbol{\xi}}(\boldsymbol{\theta}_\mathrm{q}, \boldsymbol{y}) = \begin{bmatrix} \widehat{a}_\mathrm{q}(\boldsymbol{\theta}_\mathrm{q}, \boldsymbol{y}) \\ 0 \end{bmatrix},$$

and the estimate $\widehat{\theta}_{\mathrm{o}}$ of θ is given by the update law

$$\dot{\widehat{\theta}}_{\mathrm{o}} = \gamma_{\mathrm{o}}^{-1} \boldsymbol{\psi}^{\mathrm{T}}(\widehat{\boldsymbol{y}})\widetilde{\boldsymbol{y}}. \tag{5.30}$$

An estimate of a_{q} can be obtained via an online approximator linear in the parameters [44, 52, 64]. In detail, it is assumed that a_{q} can be modeled in the following way:

$$a_{\mathrm{q}}(\boldsymbol{\theta}_{\mathrm{q}}, \boldsymbol{y}) = \sum_{i=1}^{N_\varphi} \theta_{\mathrm{q}_i} \varphi_i(y_1) + \varsigma = \boldsymbol{\theta}_{\mathrm{q}}^{\mathrm{T}} \boldsymbol{\varphi}(\boldsymbol{y}) + \varsigma, \tag{5.31}$$

where ς represents the interpolation error, $\varphi_i(y_1)$ are N_φ known basis functions, $\theta_{\mathrm{q}i}$ are unknown parameters, assumed constant (or slowly varying), and the vectors $\boldsymbol{\theta}_{\mathrm{q}}$ and $\boldsymbol{\varphi}(\boldsymbol{y})$ are defined, respectively, as

$$\boldsymbol{\theta}_{\mathrm{q}} = \begin{bmatrix} \theta_{\mathrm{q}1} \\ \theta_{\mathrm{q}2} \\ \vdots \\ \theta_{\mathrm{q}N_\varphi} \end{bmatrix}, \qquad \boldsymbol{\varphi}(\boldsymbol{y}) = \begin{bmatrix} \varphi_1(y_1) \\ \varphi_2(y_1) \\ \vdots \\ \varphi_{N_\varphi}(y_1) \end{bmatrix}. \tag{5.32}$$

When an online interpolator is used to estimate the uncertain term, the interpolation error ς can be kept bounded, provided that a suitable interpolator structure is chosen [26, 28]. Among universal approximators, Radial Basis Function Interpolators (RBFIs) provide good performance in the face of a relatively simple structure. Hence, Gaussian RBFs have been adopted, i.e.,

$$\varphi_i(y_1) = \exp\left(-\frac{|y_1 - \kappa_i|^2}{2\omega_i^2}\right), \qquad i = 1, \ldots, N_\varphi, \tag{5.33}$$

where κ_i and ω_i are the centroid and the width of the ith RBF, respectively.

The weights of the RBFI are adaptively tuned online, on the basis of the output estimation error, via the following update law:

$$\dot{\widehat{\boldsymbol{\theta}}}_{\mathrm{q}} = \gamma_{\mathrm{q}}^{-1} \boldsymbol{\varphi}(\boldsymbol{y})\widetilde{y}_1, \tag{5.34}$$

where γ_{q} is a positive gain. Then, the estimate of a_{q} is computed as follows:

$$\widehat{a}_{\mathrm{q}}(\widehat{\boldsymbol{\theta}}_{\mathrm{q}}, \boldsymbol{y}) = \widehat{\boldsymbol{\theta}}_{\mathrm{q}}^{T} \boldsymbol{\varphi}(\boldsymbol{y}). \tag{5.35}$$

In the absence of interpolation error, i.e., when $\varsigma = 0$, the convergence properties of both the state estimation error $\widetilde{\boldsymbol{x}}_{\mathrm{E}} = \boldsymbol{x}_{\mathrm{E}} - \widehat{\boldsymbol{x}}_{\mathrm{E}}$ and the parameter estimation errors, $\widetilde{\theta}_{\mathrm{o}} = \theta - \widehat{\theta}_{\mathrm{o}}$ and $\widetilde{\boldsymbol{\theta}}_{\mathrm{q}} = \boldsymbol{\theta}_{\mathrm{q}} - \widehat{\boldsymbol{\theta}}_{\mathrm{q}}$, are stated by the following theorem.

Theorem 5.2 *Under the assumption $\varsigma = 0$, there exists a set of observer gains such that the state estimation error $\widetilde{\boldsymbol{x}}_{\mathrm{E}}$ is globally uniformly convergent to $\mathbf{0}$ as $t \to \infty$, and the parameter estimation errors $\widetilde{\theta}_{\mathrm{o}}$ and $\widetilde{\boldsymbol{\theta}}_{\mathrm{q}}$ are bounded for every t.*

The proof is based on the same arguments used for Theorem 5.1 and, for the sake of completeness, is reported in Appendix A.2.

Remarks 5.1 and 5.2 can be extended to this observer. It is worth noticing that exponential convergence of the errors is of the utmost importance, since it guarantees bounded errors in the presence of bounded interpolation errors ς.

5.6.2.2 A Classical Model-Free Approach

Finally, it is worth reviewing the model-free estimation scheme proposed in [27], which is attractive for its simplicity and effectiveness.

In detail, the term a_q is considered as a further unknown parameter to be estimated, together with the heat transfer coefficient, via a suitably designed nonlinear observer [24]. The estimates $\widehat{a}_\mathrm{q}$ and $\widehat{\theta}_\mathrm{o}$ are obtained by means of the following observer:

$$\begin{bmatrix} \dot{\widehat{T}}_\mathrm{r} \\ \dot{\widehat{T}}_\mathrm{j} \\ \dot{\widehat{a}}_\mathrm{q} \\ \dot{\widehat{\theta}}_\mathrm{o} \end{bmatrix} = \begin{bmatrix} 0 & 0 & 1 & -\alpha_\mathrm{r}(T_\mathrm{r}-T_\mathrm{j}) \\ 0 & 0 & 0 & \alpha_\mathrm{j}(T_\mathrm{r}-T_\mathrm{j}) \\ 0 & 0 & 0 & 0 \\ 0 & 0 & 0 & 0 \end{bmatrix} \begin{bmatrix} \widehat{T}_\mathrm{r} \\ \widehat{T}_\mathrm{j} \\ \widehat{a}_\mathrm{q} \\ \widehat{\theta}_\mathrm{o} \end{bmatrix} + \begin{bmatrix} 0 \\ \beta_\mathrm{j}(T_\mathrm{in}-T_\mathrm{j}) \\ 0 \\ 0 \end{bmatrix}$$
$$+ \begin{bmatrix} 2l_\mathrm{q} & 0 \\ 0 & 2l_\theta \\ l_\mathrm{q}^2 & \frac{\alpha_\mathrm{r}}{\alpha_\mathrm{j}} l_\theta^2 \\ 0 & \frac{1}{\alpha_\mathrm{j}(T_\mathrm{r}-T_\mathrm{j})} l_\theta^2 \end{bmatrix} \begin{bmatrix} T_\mathrm{r}-\widehat{T}_\mathrm{r} \\ T_\mathrm{j}-\widehat{T}_\mathrm{j} \end{bmatrix}, \tag{5.36}$$

where l_q and l_θ are suitable positive gains.

It must be noticed that, when $(T_\mathrm{r}-T_\mathrm{j}) \to 0$, a singularity occurs due to the term $\frac{1}{\alpha_\mathrm{j}(T_\mathrm{r}-T_\mathrm{j})}$. Hence, in order to avoid this, $(T_\mathrm{r}-T_\mathrm{j})$ can be replaced by a constant value ϵ, when $|T_\mathrm{r}-T_\mathrm{j}| \le \epsilon$.

Convergence analysis of the observer (5.36) can be found in [24, 27].

5.7 Adaptive Two-Loop Control Scheme

The controller scheme developed in the following is based on the well-known GMC paradigm [22, 27] reviewed in Sect. 5.4.2. The key idea of this technique is that of globally linearizing the reactor dynamics by acting on the jacket temperature T_j, which is, in turn, controlled by a standard linear (e.g., PID) controller. Since T_j does not play the role of the input manipulated variable, the only way to impose an assigned behavior to the jacket temperature is that of computing a suitable setpoint $T_{\mathrm{j,des}}$, to be passed by a control loop closed around T_j. Both in [22] and [27], the mathematical relationship between the jacket temperature and the setpoint is assumed to be a known linear first-order differential equation, from which $T_{\mathrm{j,des}}$ is computed. Here, this assumption is not necessary, since a two-loop control scheme is explicitly designed.

The whole control scheme is represented in Fig. 5.2. The first control loop (inner loop) is closed around the jacket temperature in such a way to track a desired temperature, $T_{\mathrm{j,des}}(t) = y_{2,\mathrm{des}}(t)$, to be determined; then, an outer loop is closed around the reactor temperature so as to track the desired reactor temperature profile, $T_{\mathrm{r,des}}(t) = y_{1,\mathrm{des}}(t)$. The outer controller computes the desired jacket temperature on the basis of the reactor tracking error $e_1 = y_{1,\mathrm{des}} - y_1$ and of the estimate of a_{q}, while the inner controller receives $y_{2,\mathrm{des}}$ as input and computes the temperature of the fluid entering the jacket, i.e., the manipulated input u.

Since the control goal is the tracking of a temperature profile for the reactor, according to the GMC method, a desired profile for $\dot{y}_1$ must be chosen. By considering (5.12) and the expression of $\dot{y}_1 = \dot{x}_{N_{\mathrm{C}}+1}$ given by (5.15), the following equality can be imposed:

$$\widehat{a}_{\mathrm{q}} - \alpha_{\mathrm{r}}\widehat{\theta}_{\mathrm{c}}(y_1 - y_{2,\mathrm{des}}) = \dot{y}_{1,\mathrm{des}} + g_{\mathrm{P,r}}e_1 + g_{\mathrm{I,r}}\int_0^t e_1(\zeta)\,d\zeta, \tag{5.37}$$

where $\dot{y}_{1,\mathrm{des}}$ is added to take into account time-varying temperature profiles, while $g_{\mathrm{P,r}}$ and $g_{\mathrm{I,r}}$ are positive gains. In this way, it is possible to compute the desired value of the jacket temperature as

$$y_{2,\mathrm{des}} = y_1 + \frac{\dot{y}_{1,\mathrm{des}} + g_{\mathrm{P,r}}e_1 + g_{\mathrm{I,r}}\int_0^t e_1(\zeta)\,d\zeta - \widehat{a}_{\mathrm{q}}}{\alpha_{\mathrm{r}}\widehat{\theta}_{\mathrm{c}}} = y_1 + z_{2,\mathrm{des}}. \tag{5.38}$$

In the same way, by considering (5.12) and the expression of $\dot{y}_2 = \dot{x}_{N_{\mathrm{C}}+2}$ given by (5.15), a similar expression for the input u can be devised,

$$u = y_2 + \frac{\dot{y}_{2,\mathrm{des}} + g_{\mathrm{P,j}}e_2 + g_{\mathrm{I,j}}\int_0^t e_2(\zeta)\,d\zeta - \alpha_{\mathrm{j}}(y_1 - y_2)\widehat{\theta}_{\mathrm{c}}}{\beta_{\mathrm{j}}}, \tag{5.39}$$

where $e_2 = y_{2,\mathrm{des}} - y_2$ is the tracking error for the jacket temperature, and $g_{\mathrm{P,j}}$ and $g_{\mathrm{I,j}}$ are positive gains.

The parameter estimate $\widehat{\theta}_{\mathrm{c}}$ is obtained via the update law

$$\dot{\widehat{\theta}}_{\mathrm{c}} = \gamma_{\mathrm{c}}^{-1}\boldsymbol{\chi}_{\mathrm{c}}^{\mathrm{T}}(\boldsymbol{y})\boldsymbol{P}_{\mathrm{c}}\boldsymbol{\varepsilon}, \tag{5.40}$$

where γ_{c} is a positive gain, and the vectors $\boldsymbol{\varepsilon}$ and $\boldsymbol{\chi}_{\mathrm{c}}$ are defined, respectively, as

$$\boldsymbol{\varepsilon} = \begin{bmatrix} \int_0^t e_1(\zeta)\,d\zeta \\ e_1 \\ \int_0^t e_2(\zeta)d\zeta \\ e_2 \end{bmatrix}, \qquad \boldsymbol{\chi}_{\mathrm{c}}(\boldsymbol{y}) = \begin{bmatrix} 0 \\ -\alpha_{\mathrm{r}}z_{2,\mathrm{des}} \\ 0 \\ -\alpha_{\mathrm{j}}(y_1 - y_2) \end{bmatrix}. \tag{5.41}$$

The matrix $\boldsymbol{P}_{\mathrm{c}}$ in (5.40) is symmetric and positive definite,

$$\boldsymbol{P}_{\mathrm{c}} = \begin{bmatrix} \boldsymbol{P}_{\mathrm{r}} & \boldsymbol{O}_{2\times 2} \\ \boldsymbol{O}_{2\times 2} & \boldsymbol{P}_{\mathrm{j}} \end{bmatrix}, \tag{5.42}$$

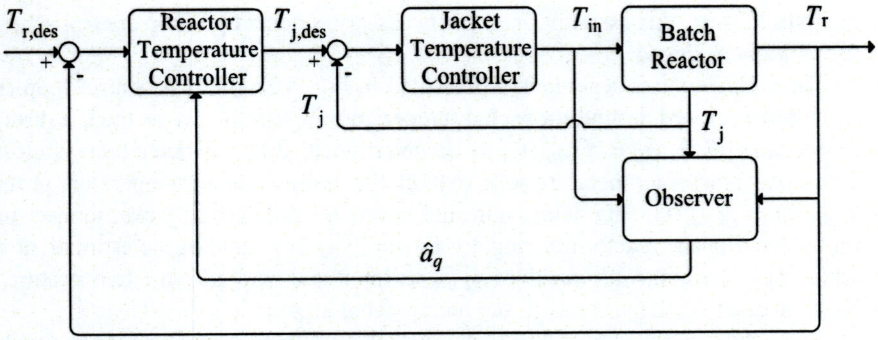

Fig. 5.2 Block scheme of the two-loop control

and each matrix $\boldsymbol{P}_*$ $(* = \mathrm{r}, \mathrm{j})$ is the (2×2) symmetric and positive definite solution of the Lyapunov equation

$$\boldsymbol{A}_*^{\mathrm{T}} \boldsymbol{P}_* + \boldsymbol{P}_* \boldsymbol{A}_* = -\boldsymbol{N}_*, \tag{5.43}$$

where

$$\boldsymbol{A}_* = \begin{bmatrix} 0 & 1 \\ -g_{\mathrm{I},*} & -g_{\mathrm{P},*} \end{bmatrix},$$

while $\boldsymbol{N}_\mathrm{r}$ and $\boldsymbol{N}_\mathrm{j}$ are symmetric positive definite matrices satisfying

$$\begin{aligned} \lambda_{\min}(\boldsymbol{N}_\mathrm{j}) &> \frac{\|\boldsymbol{P}_\mathrm{r}\|^2 \|\boldsymbol{A}_\mathrm{rj}\|^2}{\lambda_{\min}(\boldsymbol{N}_\mathrm{r})}, \\ \boldsymbol{A}_\mathrm{rj} &= \begin{bmatrix} 0 & 0 \\ 0 & \alpha_\mathrm{r}\theta \end{bmatrix}, \end{aligned} \tag{5.44}$$

and $\lambda_{\min}(\cdot)$ denotes the minimum eigenvalue of a matrix. Noticeably, since each $\boldsymbol{A}_*$ is Hurwitz (for any choice of the control gains), the solution of the Lyapunov equation (5.43) exists for any positive definite matrix $\boldsymbol{N}_*$. This implies that solutions satisfying condition (5.44) always exist.

An alternative version of the control law (5.38), (5.39) can be obtained by setting the gains $g_{\mathrm{I,r}}$ and $g_{\mathrm{I,j}}$ to zero. Therefore, the following control law can be used:

$$y_{2,\mathrm{des}} = y_1 + \frac{\dot{y}_{1,\mathrm{des}} + g_{\mathrm{P,r}} e_1 - \widehat{a}_\mathrm{q}}{\alpha_\mathrm{r} \widehat{\theta}_\mathrm{c}} = y_1 + z'_{2,\mathrm{des}}, \tag{5.45}$$

$$u = y_2 + \frac{\dot{y}_{2,\mathrm{des}} + g_{\mathrm{P,j}} e_2 - \alpha_\mathrm{j}(y_1 - y_2)\widehat{\theta}_\mathrm{c}}{\beta_\mathrm{j}}. \tag{5.46}$$

If (5.45) and (5.46) are used instead of (5.38) and (5.39), respectively, the following adaptive law can be adopted to estimate θ:

$$\dot{\widehat{\theta}}_\mathrm{c} = \gamma_\mathrm{c}^{-1} \boldsymbol{\psi}_\mathrm{c}^{\mathrm{T}}(\boldsymbol{y}) \boldsymbol{e}, \tag{5.47}$$

where γ_c is a positive gain setting the update rate of the estimate, $\boldsymbol{e}$ is the vector of the tracking errors

$$\boldsymbol{e} = \begin{bmatrix} e_1 \\ e_2 \end{bmatrix},$$

and $\boldsymbol{\psi}_c$ is the vector

$$\boldsymbol{\psi}_c(\boldsymbol{y}) = \begin{bmatrix} -\alpha_r z'_{2,des} \\ -\alpha_j(y_1 - y_2) \end{bmatrix}. \tag{5.48}$$

The presence of the integral action in the control law (5.38) and (5.39) guarantees higher robustness of the control scheme; on the other hand, tuning of this controller is more difficult since two more parameters, $g_{I,r}$ and $g_{I,j}$, are to be adjusted, and the matrix $\boldsymbol{P}_c$ satisfying (5.44) must be chosen. Moreover, even in the absence of the integral action, the performance degradation is negligible since the adaptive compensation of the parameter already includes an integral action on the tracking error. Hence, control law (5.45), (5.46) will be adopted in the case study developed in the subsequent section.

The convergence properties of the error variables (including the parameter estimation errors $\widetilde{\theta}_o = \theta - \widehat{\theta}_o$ and $\widetilde{\theta}_c = \theta - \widehat{\theta}_c$) for the overall controller–observer scheme defined by (5.23)–(5.25) and (5.38)–(5.40) are established by the following theorem:

Theorem 5.3 *If the rate constants are bounded as in* (2.32) *and* (2.33), *then, there exists a set of observer gains such that the state estimation error* $\widetilde{\boldsymbol{x}}$ *and the tracking error* $\boldsymbol{\varepsilon}$ *globally uniformly converge to* $\boldsymbol{0}$ *as* $t \to \infty$, *for any positive set of control gains. Moreover, the parameter estimation errors* $\widetilde{\theta}_o$ *and* $\widetilde{\theta}_c$ *are bounded for every* t.

The proof is based on a Lyapunov-like argument and is reported in Appendix A.3.

As for the overall closed-loop system resulting defined by (5.23)–(5.25) and (5.45)–(5.47), the following theorem holds:

Theorem 5.4 *If the rate constants are bounded as in* (2.32) *and* (2.33), *then, there exists a set of observer gains such that the state estimation error* $\widetilde{\boldsymbol{x}}$ *and the tracking error* $\boldsymbol{e}$ *globally uniformly converge to* $\boldsymbol{0}$ *as* $t \to \infty$, *for any positive set of control gains. Moreover, the parameter estimation errors* $\widetilde{\theta}_o$ *and* $\widetilde{\theta}_c$ *are bounded for every* t.

The proof is based on the same arguments used in the proof of Theorem 5.3 and is reported, for the sake of completeness, in Appendix A.4.

Remarks 5.1 and 5.2 on the exponential stability of the estimation error dynamics can be extended to the overall controller–observer scheme as well. Hence, robustness with respect to effects due to modeling uncertainties and/or disturbances is guaranteed. Moreover, the following remarks can be stated.

Remark 5.3 Although the stability analysis considers the dynamics of the overall system (i.e., the dynamics of both the observer and the controller), tuning of the observer gains ($\boldsymbol{L}$ and γ_o) and of the controller gains ($g_{P,r}$, $g_{P,j}$, and γ_c and, eventually,

$g_{\mathrm{I,r}}$, $g_{\mathrm{I,j}}$, and P_{c}) can be achieved separately, since the stability conditions do not put mutual constraints on the two set of gains.

Remark 5.4 As can be noticed from (5.40) and (5.47), the estimate, $\widehat{\theta}_{\mathrm{c}}$, used in the control law is different from the estimate $\widehat{\theta}_{\mathrm{o}}$ computed by the observer. This is necessary since the convergence of the latter to θ is not guaranteed unless the persistency of excitation condition is fulfilled. Hence, a different update law is adopted to ensure the convergence of the controller tracking errors. A unique update law could be adopted, ensuring the convergence of both estimation and tracking errors. In this case, however, stability conditions will put mutual constraints on the observer and controller parameters, and thus independent tuning of the two structures would not guarantee the convergence of the estimation/tracking errors.

Remark 5.5 The stability analysis has been developed for the controller in conjunction with the observer (5.23), (5.24). It can be verified that, if the model-free observer defined by (5.29), (5.30), and (5.34) is considered, the stability of the overall scheme can be proven by using similar arguments.

5.8 Case Study: Temperature Control

In this section, the proposed approach has been tested in a simulation case study, developed in the MATLAB/SIMULINK© environment. In detail, the problem of temperature control of the phenol–formaldehyde reactive system, developed in Sect. 2.4, has been considered.

In order to analyze the effects of different estimates of a_{q} on the control scheme, the cascade controller (5.45)–(5.47) is used in conjunction with the three different observers presented in Sect. 5.6. Finally, the results are compared with those obtained by using the most widely used industrial controller, i.e., the PID controller. Therefore, five different control schemes have been compared:

- *Model-free scheme*: the two-loop scheme (5.45)–(5.47) is adopted in conjunction with the model-free observer (5.36), providing $\widehat{a}_{\mathrm{q}}$ and $\widehat{\theta}_{\mathrm{o}}$. In order to cope with the singularities occurring when $(T_{\mathrm{r}} - T_{\mathrm{j}}) \to 0$, this term has been replaced by the constant value 0.1 when $|T_{\mathrm{r}} - T_{\mathrm{j}}| \leq 0.1$.
- *RBFI-based scheme*: the two-loop scheme (5.45)–(5.47) is adopted in conjunction with the RBFI-based model-free observer defined by (5.29), (5.30), (5.34), and (5.35). The RBFI uses 15 Radial Basis Functions; the centroids are chosen evenly distributed in the interval [293 K, 368 K], considered as the range of temperatures of the reaction; the width, the same for all the RBFs, has been set to 10^2.
- *Model-based scheme*: the model-based controller–observer scheme defined by (5.23) and (5.45), (5.46) is adopted, where the adaptive estimation is not introduced, and thus the available nominal estimate of θ is used.
- *Adaptive model-based scheme*: the adaptive model-based controller–observer scheme defined by (5.23)–(5.25) and (5.45)–(5.47) is adopted.

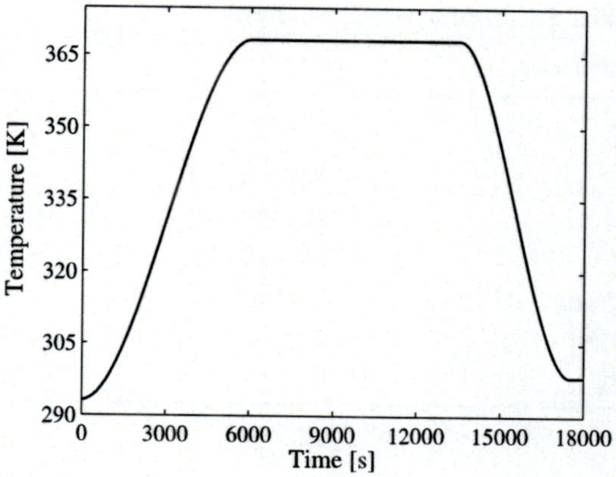

Fig. 5.3 Desired reactor temperature profile

- *PID controller*: the classical PID regulator (5.1) with a feedforward compensation of the desired reactor temperature (i.e., $y_{1,\text{des}}$ is added to the control input).

In order to perform a fair comparison between the above-mentioned control strategies, all the schemes have been tuned so as to achieve the same control effort (i.e., so as to obtain the same time histories of u, as far as possible).

A realistic temperature profile $T_{\text{r,des}}(t) = y_{1,\text{des}}(t)$, reported in Fig. 5.3, has been chosen as desired reactor temperature; it is characterized by three phases:

- *heating phase*: in this phase the desired reactor temperature is raised from its initial value, 293 K, to 368 K in 6000 s, via a third-order polynomial with null initial and final derivatives
- *isothermal phase*: in this phase a constant temperature setpoint (368 K) is commanded for 7500 s; and
- *cooling phase*: in this phase the desired temperature is driven to 298 K in 4000 s; the profile is a third-order polynomial with null initial and final derivatives. The final temperature is then kept constant for 500 s.

5.8.1 Simulation Model

The phenol–formaldehyde reaction has been simulated by considering the kinetic model containing 13 components and 89 reactions developed in Sect. 2.4. A well-mixed batch reactor, equipped with a cooling/heating jacket fed with liquid water, has been considered. The mass balances have been written according to the general scheme (2.27), while the energy balances are written as in (2.30), (2.31). The values of the main model parameters are reported in Table 5.1, where the heat capacity of the reacting solution has been taken from [50]. The temperature of the water entering the jacket ranges from $T_{\text{in,min}} = 285$ K to $T_{\text{in,max}} = 370$ K, while the volumetric flow rate is fixed at $F_{\text{V}} = 0.1\ \text{m}^3\ \text{s}^{-1}$.

In order to simulate a realistic industrial reactor, some important assumptions on the actuator and the sensors have been done. Namely, a first-order linear dynamics,

Table 5.1 Main simulation parameters

Parameter	Value	Parameter	Value
V_r	$6\ \mathrm{m}^3$	V_j	$1.729\ \mathrm{m}^3$
θ	$11.5\ \mathrm{kJ\,s^{-1}\,K^{-1}}$	F_V	$0.1\ \mathrm{m^3\,s^{-1}}$
ρ_r	$1.0 \cdot 10^3\ \mathrm{kg\,m^{-3}}$	ρ_j	$1.0 \cdot 10^3\ \mathrm{kg\,m^{-3}}$
c_r	$1.712\ \mathrm{kJ\,kg^{-1}\,K^{-1}}$	c_j	$4.186\ \mathrm{kJ\,kg^{-1}\,K^{-1}}$
$C_{Ph}(0)$	$4200\ \mathrm{mol\,m^{-3}}$	$C_F(0)$	$7560\ \mathrm{mol\,m^{-3}}$
$T_r(0)$	293.15 K	$T_j(0)$	293.16 K
$T_{in,min}$	285 K	$T_{in,max}$	370 K

with a time constant of 3 s, has been introduced in the simulation model (but not considered in the controller) between the commanded control input (computed by the controller) and the real temperature of the water entering the jacket. Moreover, it has been assumed that only the reactor and the jacket temperatures are measured; Gaussian white noise with zero mean and variance of $5 \cdot 10^{-3}\ \mathrm{K}^2$ is added to temperature measurements.

Initial conditions for the reactant concentrations and the temperatures of the vessel and the jacket are reported in Table 5.1.

An initial estimation error on the concentrations, which corresponds to 5% of their true values, has been assumed. Moreover, a wrong nominal estimate of θ has been considered, which is assumed to be equal to 1.6 times its true value (i.e., a 60% error has been introduced).

5.8.2 Design of the Controller–Observer Scheme

The model-based controller–observer scheme requires to solve online the system of differential equations of the observer. The phenol–formaldehyde reaction model is characterized by 15 differential equations, and it is, thus, unsuitable for online computations. To overcome this problem, one of the reduced models developed in Sect. 3.8.1 can be adopted. In order to be consistent with the general form of non-chain reactions (2.27) adopted to develop the controller–observer scheme, the reduced model (3.57) with first-order kinetics has been used to design the observer. The mass balances of the reduced model are given by

$$\begin{cases} \dot{C}_A = -k_{c1}(T_r)C_A, \\ \dot{C}_I = k_{c1}(T_r)C_A - k_{c2}(T_r)C_I, \\ \dot{C}_P = k_{c2}(T_r)C_I - k_{c3}(T_r)C_P, \end{cases} \tag{5.49}$$

while the effect of the heat released by the reaction on the reactor temperature is computed as

$$a_q = a_1(T_r)C_A + a_2(T_r)C_I + a_3(T_r)C_P, \tag{5.50}$$

Table 5.2 Observer gains for the control schemes compared in simulation

Gains	Model-free	RBFI-based	Model-based	Adaptive model-based
l_1, l_2, l_3			$10, 5 \cdot 10^2, 10^{-1}$	$10, 5 \cdot 10^2, 10^{-1}$
$l_\mathrm{r}, l_\mathrm{j}$		1, 1	1, 1	1, 1
l_q, l_θ	$2 \cdot 10^{-2}, 6.5 \cdot 10^{-3}$			
γ_o		$1.5 \cdot 10^{-1}$		$1.5 \cdot 10^{-1}$
γ_q		$3 \cdot 10^{-2}$		

where

$$a_i(T_\mathrm{r}) = \frac{-\Delta H_{\mathrm{R}i}(T_\mathrm{r}) k_{\mathrm{c}i}(T_\mathrm{r})}{\rho_\mathrm{r} c_\mathrm{r}}, \quad i = 1, 2, 3.$$

The values of $k_{\mathrm{c}i}(T_\mathrm{r})$ and $\Delta H_{\mathrm{R}i}(T_\mathrm{r})$ $(i = 1, \dots, 3)$ have been identified in Sect. 3.8.

The state vector can be defined as

$$\boldsymbol{x} = \begin{bmatrix} x_1 \\ x_2 \\ x_3 \\ x_4 \\ x_5 \end{bmatrix} = \begin{bmatrix} C_\mathrm{A} \\ C_\mathrm{I} \\ C_\mathrm{P} \\ T_\mathrm{r} \\ T_\mathrm{j} \end{bmatrix}. \tag{5.51}$$

The matrix $\boldsymbol{A}(\boldsymbol{y})$ of the observer (5.23) has the form

$$\boldsymbol{A}(\boldsymbol{y}) = \begin{bmatrix} -k_{\mathrm{c}1}(T_\mathrm{r}) & 0 & 0 & 0 & 0 \\ k_{\mathrm{c}1}(T_\mathrm{r}) & -k_{\mathrm{c}2}(T_\mathrm{r}) & 0 & 0 & 0 \\ 0 & k_{\mathrm{c}2}(T_\mathrm{r}) & -k_{\mathrm{c}3}(T_\mathrm{r}) & 0 & 0 \\ a_1(T_\mathrm{r}) & a_2(T_\mathrm{r}) & a_3(T_\mathrm{r}) & 0 & 0 \\ 0 & 0 & 0 & 0 & 0 \end{bmatrix}. \tag{5.52}$$

The model-based observer requires tuning of 6 parameters, i.e., the nonzero values in matrix $\boldsymbol{L}$ and γ_o. As for the model-free observer defined by (5.29), (5.30), and (5.34), the dynamics of the reaction is not required, and only two gains (the main diagonal of matrix $\boldsymbol{L}_\mathrm{E}$) and two update gains ($\gamma_\mathrm{o}$ and γ_q) are needed. Finally, the observer (5.36) requires tuning of the two gains l_θ and l_q. All the above gains have been tuned via a trial-and-error procedure and are summarized in Table 5.2.

As stated in Remark 5.3, the parameters of the two-loop controller ($g_{\mathrm{P,r}}, g_{\mathrm{P,j}}$) and γ_c have been chosen independently from the adopted observer via a trial-and-error procedure. Therefore, they are identical for all the control schemes and are reported in Table 5.3. In the controller–observer scheme without adaption (i.e., using the nominal estimate of θ), both γ_o and γ_c have been set to zero.

5.8.3 Discussion of Results

Figures 5.4 to 5.7 show the simulation results. It can be recognized that satisfactory temperature tracking performance (Fig. 5.4) are achieved in all cases, with similar

Table 5.3 Two-loop controller gains

$g_{P,r}$	$g_{P,j}$	γ_c
$5 \cdot 10^{-2}$	$5 \cdot 10^{-2}$	$6 \cdot 10^{-2}$

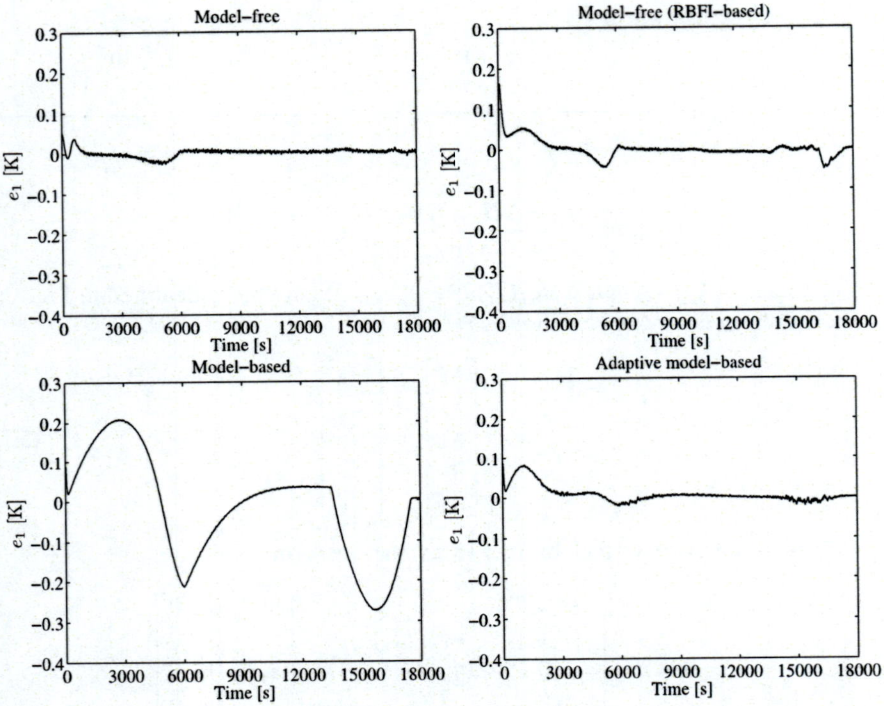

Fig. 5.4 Reactor temperature tracking errors

values of the control input (Fig. 5.5). Noticeably, the model-based adaptive scheme achieves tracking performance comparable with respect to the model-free schemes, also in the presence of large model uncertainties. Moreover, the adaptive approaches outperform the nonadaptive model-based scheme, since the latter does not take into account the parametric uncertainties at all.

It can be argued that the differences between the compared schemes are mainly due to the different estimation accuracy of the quantity a_q (Fig. 5.6). It can be seen that, after the initial transient phase in which the model-free observers present an *inverse response*, both the adaptive (model-based and model-free) approaches achieve very good estimates. As for the parameter estimate, since both the adaptive observers ($\widehat{\theta}_o$) and the controller ($\widehat{\theta}_c$) estimates converge to the true value of θ (see Fig. 5.7), it is possible to argue that the persistency of excitation condition is fulfilled.

As expected, the effect of the singularity in the model-free observer (5.36) is visible in the last part of the batch, where T_r and T_j tends to be equal, and thus the

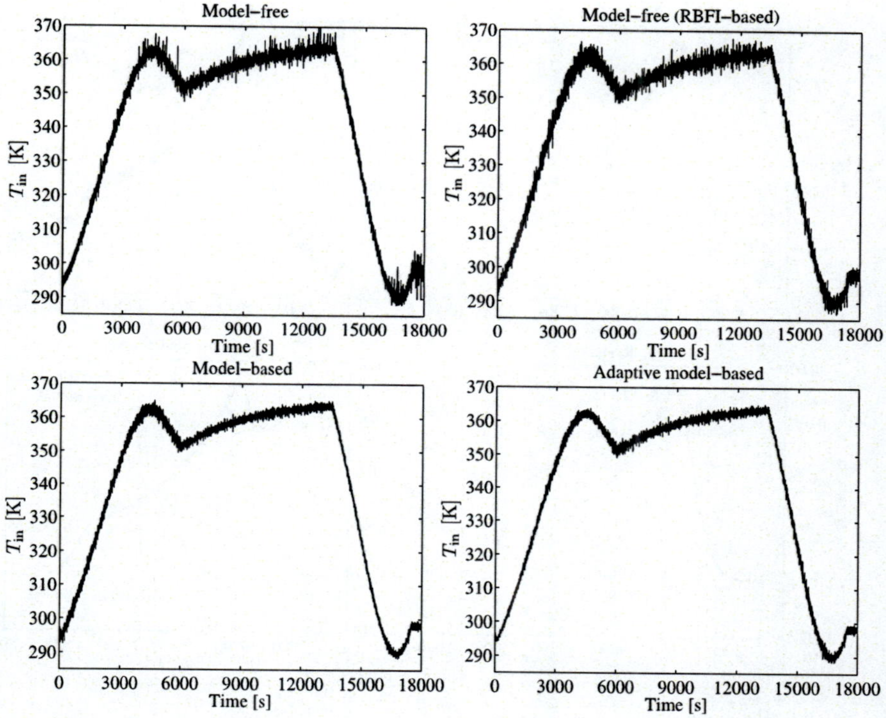

Fig. 5.5 Commanded temperature of the fluid entering the jacket

estimate deviates from the true value of the parameter. The singularity, however, does not influence the overall behavior of the closed-loop system, since it occurs during the final stage of the reactivity of the system.

The effect of measurement noise can be observed for all variables (especially the control input, as shown in Fig. 5.5). In particular, the model-free observer based on RBFI is the most sensitive to measurement noise.

It can be concluded that the exponential stability property confers to the adaptive model-based scheme a satisfactory degree of robustness. Therefore, even in the presence of large model uncertainties, its performance is comparable with or better than that of model-free approaches.

5.8.4 Comparison with the PID Controller

The following values have been adopted for the PID gains: $g_P = 10$, $g_I = 5 \cdot 10^{-1}$, and $g_D = 10^{-3}$. In order to perform a fair comparison between the PID approach and the two-loop controller-observer strategies, the PID gains have been tuned via a trial-and-error procedure so as to achieve the same control effort of the other considered control strategies (i.e., so as to obtain the same time histories of u as far as possible).

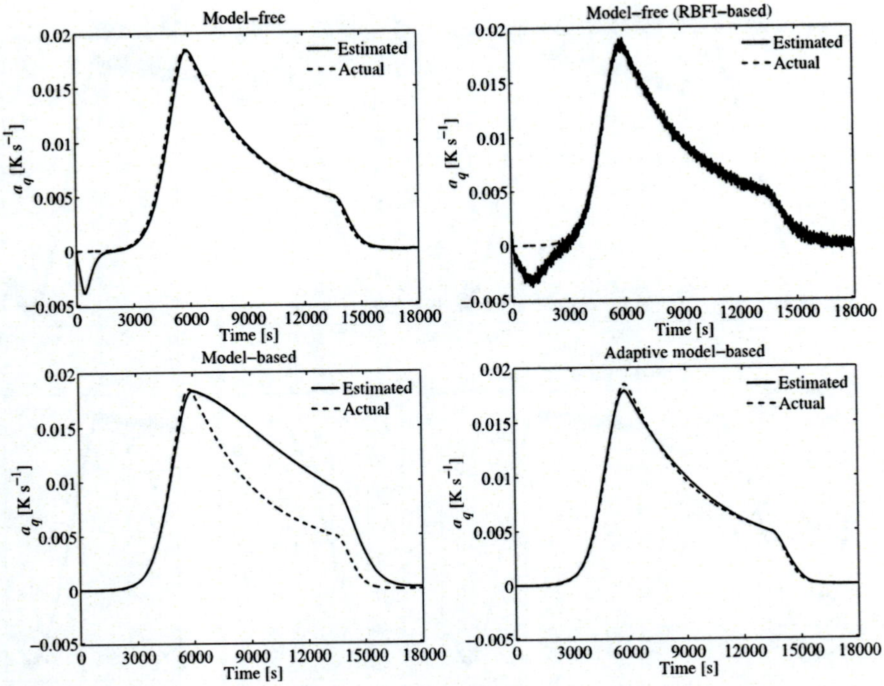

Fig. 5.6 Estimates of a_q

Figure 5.8 is referred to the performance of the PID controller. It can be recognized that the temperature tracking error is always below 0.5 K and is characterized by a very similar time history of the control input with respect to the previously tested schemes (Fig. 5.5). By comparing the results in Fig. 5.8 with those in Fig. 5.4, the adaptive model-based scheme presents better performance than the linear PID control, especially during the heating and cooling phases, i.e., when the reference temperature is not constant.

The major advantages deriving from the application of the model-based two-loop control strategies can be highlighted in terms of their ability to prevent potentially dangerous conditions in the reactor. As already discussed in Sect. 4.6, the safe operative conditions for this reactive system are to be determined—at least in the early reaction stages—in the light of a maximum temperature limit imposed by the bubbling point of the liquid, rather than by classical runaway boundaries determined by geometric or sensitivity-based criteria. By assuming a precautionary maximum temperature of 371 K (98 °C), the performance of the two control strategies are evaluated by determining the critical value of $\theta = US$, defined as the lowest value that generates a temperature–time profile in the reactor with maximum lower than 371 K.

The results are shown in Fig. 5.9 as a function of the set-point temperature for both the model-based and the PID controller. In both cases, the critical θ increases with the setpoint temperature since a more efficient heat exchange is required when

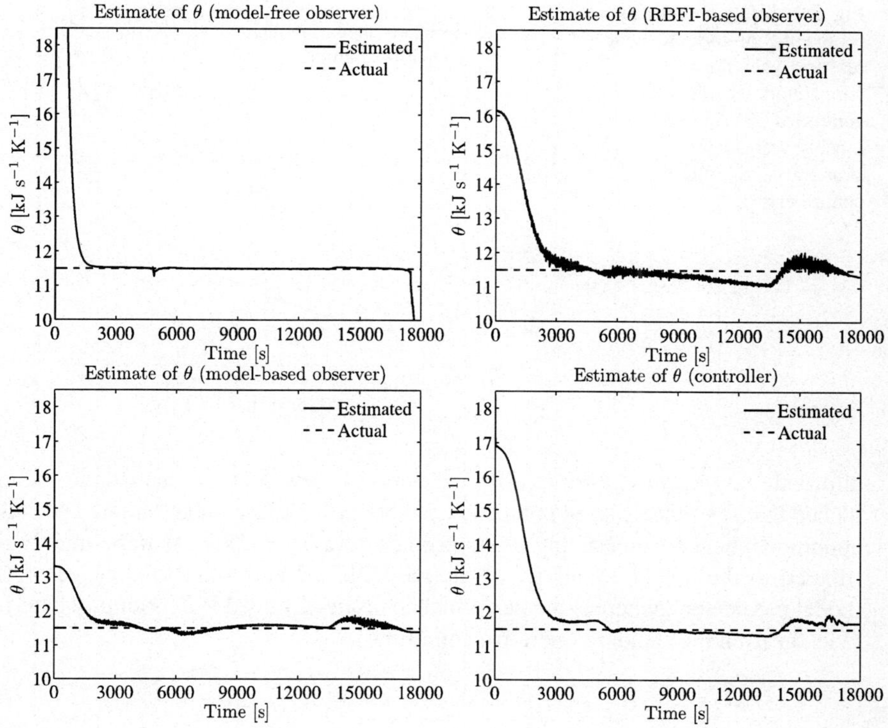

Fig. 5.7 Estimates of the parameter θ

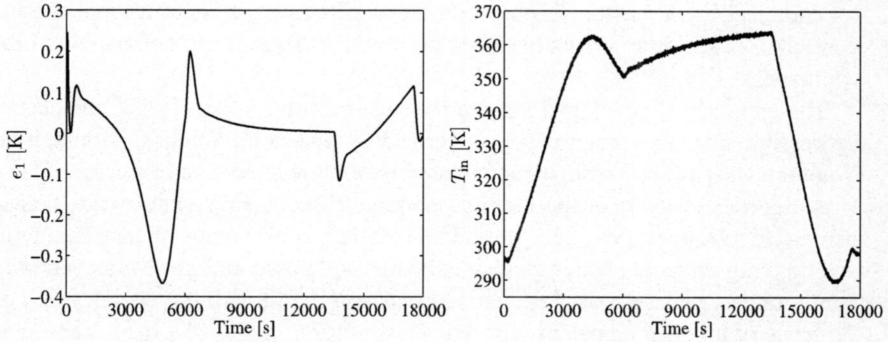

Fig. 5.8 Reactor temperature tracking error (*left*) and commanded temperature of the fluid in the jacket (*right*) obtained by using PID controller

working closer to the imposed limit; a vertical asymptote at the maximum allowable temperature of 371 K is present in both cases. Interestingly, the critical θ values for the PID-controlled reactor prove to be considerably higher than those obtained with the model-based approach; hence, in the latter case a much lower heat exchange area and/or coefficient are required in order to prevent overcoming of the maximum

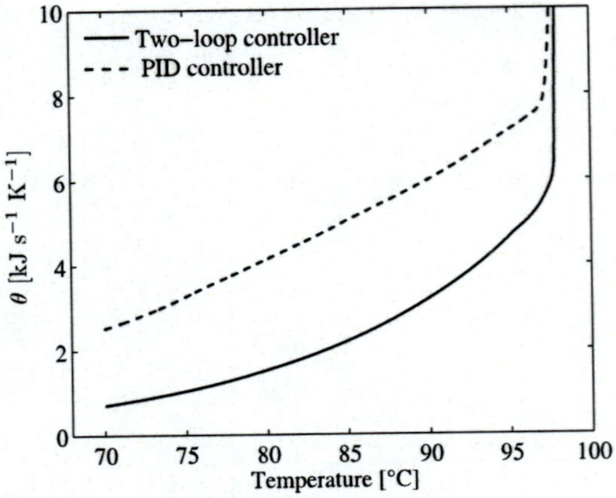

Fig. 5.9 Safety boundaries for the phenol–formaldehyde reaction when the temperature reactor is controlled by using the model-based adaptive strategy and the PID controller

allowable temperature. From a different point of view, this corresponds to the evidence that the model-based controlled reactor has a wider range of safe operative conditions, hence emphasizing its superior performance. This has to be mainly attributed to the higher adaptivity of the model-based controller to changes of the model parameters, whereas the performance of a well-tuned PID promptly deteriorates far from the standard operative conditions.

5.9 Conclusions

In this chapter an overview of the most widely adopted temperature control schemes for chemical batch reactors has been provided. Moreover, an adaptive model-based controller–observer approach has been proposed, analyzed, and compared to other approaches.

The approach is developed for a fairly wide class of processes, i.e., the class of irreversible nonchain reactions characterized by first-order kinetics. Although this is not the most general case, it encompasses several real reactive processes.

A rigorous analysis of the main properties of the overall scheme (i.e., convergence and robustness) has been provided. In detail, convergence of state estimation and tracking errors is always guaranteed under mild assumptions. Moreover, when the (stronger) persistency of excitation condition is fulfilled, the exponential convergence of all error signals is ensured. This, in turn, implies the robustness of the proposed scheme in the face of uncertainties and disturbances.

Moreover, it is worth remarking that the use of an accurate and reliable state observer, which is necessary for the proposed controller, can be advantageous for other purposes as well (e.g., process monitoring and fault diagnosis).

Since the design and the tuning of the observer can be achieved independently from the adopted controller, the latter can be adopted in conjunction with observers different from those presented in this chapter. The model-based observer (5.23) needs a good knowledge of the reaction kinetics: this may be regarded as a limitation for its practical application, where a certain degree of mismatch between the modeled and the real reaction mechanism is always present. Nevertheless, in the presence of bounded and/or vanishing uncertainties, the property of exponential convergence ensures a certain degree of robustness of the controller–observer scheme. In other words, if the mismatch between the model and the real kinetics is bounded (vanishing), bounded (asymptotically convergent) estimation and tracking errors are expected. Of course, modeling errors must be kept as small as possible, via suitable modeling and identification techniques of the reaction dynamics.

When an accurate model of the reaction kinetics is not available, the estimation of the heat released can be achieved by resorting to model-free adaptive approaches, where the estimation of both unknown parameters and the effect of the heat released by the reaction can be computed by resorting, e.g., to universal interpolators.

In the case study, the adaptive model-based approach is designed on the basis of a reduced model of the phenol–formaldehyde reaction introduced in the Chap. 2. Noticeably, the results show that the model-based control scheme achieves very good performance even when a strongly simplified mathematical model of the reactive system is adopted for the design.

References

1. P.S. Agachi, Z.K. Nagy, M.V. Cristea, and A. Imre-Lucaci. *Model Based Control. Case Studies in Process Engineering*. Wiley-VCH, Weinheim, 2006.
2. M. Agarwal. Combining neural and conventional paradigms for modeling, prediction and control. *International Journal of Systems Science*, 28:65–81, 1997.
3. A. Altınten, F. Ketevanlioğlu, S. Erdoğan, H. Hapoğlu, and M. Alpbaz. Self-tuning PID control of jacketed batch polystyrene reactor using genetic algorithm. *Chemical Engineering Journal*, 138:490–497, 2008.
4. J. Alvarez-Ramirez and J. Alvarez. Robust temperature control for batch chemical reactors. *Chemical Engineering Science*, 60:7108–7120, 2005.
5. K.J. Aström and B. Wittenmark. *Adaptive Control, 2nd Edition*. Addison-Wesley, Reading, 1995.
6. N. Aziz, M.A. Hussain, and I.M. Mujtaba. Performance of different types of controllers in tracking optimal temperature profiles in batch reactors. *Computers and Chemical Engineering*, 24:1069–1075, 2000.
7. J.B. Balchen, B. Lie, and I. Solberg. Internal decoupling in nonlinear process control. *Modeling Identification and Control*, 9:137–148, 1988.
8. R.D. Bartusiak, C. Georgakis, and M.J. Reilly. Nonlinear feedforward/feedback control structures designed by reference synthesis. *Chemical Engineering Science*, 44:1837–1851, 1989.
9. V.M. Becerra, P.D. Roberts, and G.W. Griffiths. Applying the extended Kalman filter to systems described by nonlinear differential-algebraic equations. *Control Engineering Practice*, 9:267–281, 2001.
10. M.A. Beyer, W. Grote, and G. Reinig. Adaptive exact linearization control of batch polymerization reactors using a Sigma-Point Kalman filter. *Journal of Process Control*, 18:663–675, 2008.
11. N.V. Bhat and T.J. McAvoy. Use of neural nets for dynamic modelling and control of chemical process systems. *Computers and Chemical Engineering*, 14:583–599, 1990.
12. S.P. Bhattacharyya, A. Datta, and L.H. Keel. *Linear Control Theory*. CRC Press, Boca Raton, 2009.
13. D. Bonvin, P. de Valliére, and D.W.T. Rippin. Application of estimation techniques to batch reactors. I. Modelling thermal effects. *Computers and Chemical Engineering*, 13:1–9, 1989.

14. M.W. Brown, P.L. Lee, G.R. Sullivan, and W. Zhou. A constrained nonlinear multivariable control algorithm. *Transactions IChemE*, 68:464–476, 1990.
15. F. Caccavale, M. Iamarino, F. Pierri, and V. Tufano. An adaptive controller-observer scheme for temperature control of non-chain reactions in batch reactors. *International Journal of Adaptive Control and Signal Processing*, 22:627–651, 2008.
16. F. Caccavale, F. Pierri, and L. Villani. An adaptive observer for fault diagnosis in nonlinear discrete-time systems. *ASME Journal of Dynamic Systems, Measurement and Control*, 130:1–9, 2008.
17. F. Cameron and D.E. Seborg. A self-tuning controller with a PID structure. *International Journal of Control*, 38:401, 1983.
18. T. Clark-Pringle and J.F. MacGregor. Nonlinear adaptive temperature of multi-product semi-batch polimerization reactors. *Computers and Chemical Engineering*, 21:1395–1405, 1997.
19. D.W. Clarke and C. Mohtadi. Properties of generalized predictive control. *Automatica*, 25:859–875, 1989.
20. D.W. Clarke, C. Mohtadi, and P.S. Tuffs. Generalized predictive control—Part I. The basic algorithm. *Automatica*, 23:137–148, 1987.
21. D.W. Clarke, C. Mohtadi, and P.S. Tuffs. Generalized predictive control—Part II. Extensions and interpretations. *Automatica*, 23:149–160, 1987.
22. B.J. Cott and S. Macchietto. Temperature control of exothermic batch reactors using generic model control (GMC). *Industrial & Engineering Chemistry Research*, 28:1177–1184, 1989.
23. J.R. Cutler and B.L. Ramaker. Dynamic matrix control—a computer control algorithm. In *Proceedings of the Joint Automatic Control Conference*, 1980.
24. M. Farza, K. Busawon, and H. Hammouri. Simple nonlinear observers for on-line estimation of kinetic rates in bioreactors. *Automatica*, 34:301–318, 1998.
25. C. Filippi, J.L. Greffe, J. Bordet, J. Villermaux, J.L. Barnay, B. Ponte, and C. Georgakis. Tendency modeling of semi-batch reactors for optimization and control. *Chemical Engineering Science*, 41:913–920, 1986.
26. K. Funahashi. On the approximate realization of continuous mappings by neural networks. *Neural Networks*, 2:183–192, 1989.
27. B. Guo, A. Jiang, X. Hua, and A. Jutan. Nonlinear adaptive control for multivariable chemical processes. *Chemical Engineering Science*, 56:6781–6791, 2001.
28. S. Haykin. *Neural Networks: A Comprehensive Foundation*. Prentice Hall, Upper Saddle River, 1998.
29. M.A. Henson and D.E. Seborg. An internal model control strategy for nonlinear systems. *AIChE Journal*, 37:1065–1081, 1991.
30. M.A. Henson and D.E. Seborg. Adaptive nonlinear control of ph neutralization process. *IEEE Transactions on Control Systems Technology*, 2:169–179, 1994.
31. M.A. Henson and D.E. Seborg. *Nonlinear Process Control*. Prentice Hall, Upper Saddle River, 1997.
32. G. Karer, I. Škrjanc, and B. Zupančič. Self-adaptive predictive functional control of the temperature in an exothermic batch reactor. *Chemical Engineering & Processing: Process Intensification*, 47:2379–2385, 2008.
33. E. Katende and A. Jutan. A new constrained self-tuning PID controller. *The Canadian Journal of Chemical Engineering*, 71:625–633, 1993.
34. E. Katende and A. Jutan. Nonlinear predictive control of complex processes. *Industrial and Engineering Chemistry*, 35:2721–2728, 1996.
35. H.K. Khalil. *Nonlinear Systems, 2nd Edition*. Prentice Hall, Upper Saddle River, 1996.
36. K. Konakom, P. Kittisupakorn, and I.M. Mujtaba. Batch control improvement by model predictive control based on multiple reduced-models. *Chemical Engineering Journal*, 145:129–134, 2008.
37. C. Kravaris and C.B. Chung. Nonlinear state feedback synthesis by global input–output linearization. *AIChE Journal*, 33:592–603, 1987.
38. J.H. Lee, M. Morari, and C.E. Garcia. State-space interpretation of model predictive control. *Automatica*, 30:707–717, 1994.

39. K.S. Lee and J.H. Lee. Model predictive control for non linear batch processes with asymptotically perfect tracking. *Computers and Chemical Engineering*, 21:873–879, 1997.
40. P.L. Lee and R.B. Newell. Generic model control—a case study. *Canadian Journal of Chemical Engineering*, 67:478–483, 1989.
41. P.L. Lee and G.R. Sullivan. Generic model control. *Computers and Chemical Engineering*, 12:573–580, 1988.
42. P.L. Lee and W. Zhou. A new multivariable deadtime control algorithm. *Chemical Engineering Communications*, 91(1):49–63, 1990.
43. S. Li, K.J. Lim, and D.G. Fisher. A state-space formulation of model predictive control. *AIChE Journal*, 35:241–249, 1989.
44. L. Ljung, J. Sjoberg, and H. Hjalmarsson. On neural networks model structures in system identification. In S. Bittanti and G. Picci, editors, *Identification, Adaptation, Learning*, pages 366–393. Springer, Berlin, 1996.
45. R. Luus and O.N. Okongwu. Towards practical optimal control of batch reactors. *Chemical Engineering Journal*, 75:1–9, 1999.
46. L. Magni, D.M. Raimondo, and F. Allgöwer, editors, *Nonlinear Model Predictive Control.* Springer, Berlin, 2009.
47. I.M. Mujtaba, N. Aziz, and M.A. Hussain. Neural network based modelling and control in batch reactor. *Chemical Engineering Research and Design*, 84:635–644, 2006.
48. Z.K. Nagy and R.D. Braatz. Robust nonlinear model predictive control of batch processes. *AIChE Journal*, 49:1776–1786, 2003.
49. A.A. Patwardhan, J.B. Rawlings, and T.F. Edgar. Nonlinear model predictive control. *Chemical Engineering Communications*, 87:1–23, 1990.
50. I. Pektas. High-temperature degradation of reinforced phenolic insulator. *Journal of Applied Polymer Science*, 67:1877–1883, 1998.
51. F. Pierri, G. Paviglianiti, F. Caccavale, and M. Mattei. Observer-based sensor fault detection and isolationfor chemical batch reactors. *Engineering Applications of Artificial Intelligence*, 21:1204–1216, 2008.
52. M.M. Polycarpou and A.J. Helmicki. Automated fault detection and accommodation: a learning systems approach. *IEEE Transactions on Systems, Man, and Cybernetics*, 25:1447–1458, 1995.
53. J. Richalet, A. Rault, J.L. Testud, and J. Papon. Model predictive heuristic control: application to industrial processes. *Automatica*, 14:413–428, 1978.
54. H. Schuler and C.U. Schmidt. Calorimetric state estimators for chemical reactor diagnosis and control: reviews of methods and application. *Chemical Engineering Science*, 47(4):899–915, 1992.
55. H. Seki, M. Ogawa, S. Ooyama, K. Akamatsu, M. Ohshima, and W. Yang. Industrial application of a nonlinear model predictive control to polymerization reactors. *Control Engineering Practice*, 9:819–828, 2001.
56. P.D. Signal and P.L. Lee. Generic model adaptive control. *Chemical Engineering Communications*, 115(1):35–52, 1992.
57. P. Tatjewski. *Advanced Control of Industrial Processes*. Springer, London, 2007.
58. O. Ubricha, B. Srinivasanb, P. Lerenac, D. Bonvin, and B. Stoessel. The use of calorimetry for on-line optimisation of isothermal semi-batch reactors. *Chemical Engineering Science*, 56:5147–5156, 2001.
59. P. Vega, P. Prada, and V. Aleixandre. Self-tuning predictive PID controller. In *Proceeding of Inst. Electrical Engineering*, pages 303–308, 1991.
60. D. Wang, D.H. Zhou, Y.H. Jin, and S.J. Qin. A strong tracking predictor for nonlinear processes with input time delay. *Computers and Chemical Engineering*, 28:2523–2540, 2004.
61. D.I. Wilson, M. Agarwal, and D.V.T. Rippin. Experiences implementing the extended Kalman filter on an industrial batch reactor. *Combustion and Flame*, 22:1653–1672, 1998.
62. X.Q. Xie, D.H. Zhou, and Y.H. Jin. Strong tracking filter based adaptive generic model control. *Journal of Process Control*, 9:337–350, 1999.

63. K. Yamuna Rani and S.C. Patwardhan. Data-driven model based control of a multi-product semi-batch polymerization reactor. *Chemical Engineering Research and Design*, 85:1397–1406, 2007.
64. X. Zhang, M.M. Polycarpou, and T. Parisini. A robust detection and isolation scheme for abrupt and incipient faults in nonlinear systems. *IEEE Transactions on Automatic Control*, 47:576–593, 2002.
65. W. Zhou and P.L. Lee. Robust stability analysis of generic model control. *Chemical Engineering Communications*, 117(1):41–72, 1992.

Chapter 6
Fault Diagnosis

List of Principal Symbols

$A_d(y)$	matrix defined in (6.6)
f	fault vector
$\mathcal{F}_a$	admissible set of actuator/process faults
g, h	functions defining the state-space model in (6.1)
k_c	rate constant $[(\text{mol/m}^{-3})^{1-n}\ \text{s}^{-1}]$
L	matrix gains of the observers
n	vector of measurement noise
N_C	number of compounds involved in the reaction
N_F	number of considered actuator/process faults
r	scalar residual
$\boldsymbol{r}$	residual vector
$S_{j,i}$	temperature sensors in the cooling jacket ($i = 1, 2$)
$S_{r,i}$	temperature sensors in the reactor ($i = 1, 2$)
t	time [s]
T	temperature [K]
t_f	fault time [s]
$\mathcal{T}$	time set
u	control input variable
U	overall heat transfer coefficient $[\text{J m}^{-2}\ \text{K}^{-1}\ \text{s}^{-1}]$
$\mathcal{U}$	set of admissible inputs vectors
x	vector of state variables
$\mathcal{X}$	set of admissible state vectors
y	vector of measured output variables
$\mathcal{Y}$	set of admissible output vectors
$\lVert \cdot \rVert$	Euclidean norm

Greek Symbols

γ	positive gain setting the parameter estimate update rate
δ	magnitude of the fault
η	vector of system uncertainties

F. Caccavale et al., *Control and Monitoring of Chemical Batch Reactors*,
Advances in Industrial Control,
DOI 10.1007/978-0-85729-195-0_6,

θ	parameter US
$\boldsymbol{\theta}_{\mathrm{f}}$	vector of unknown parameters characterizing the fault magnitude
μ	normalization factor of residuals
τ	time constant setting the fault evolution rate
$\boldsymbol{\varphi}$	regressor matrix of the fault model

Subscripts and Superscripts

a	actuator
E	energy balance
f	fault
j	jacket
m	measured
M	mass balance
max	maximum
min	minimum
p	process
r	reactor
s	sensor
SM	variables referred to the observers SM1 and SM2
u	fault affecting the cooling system
U	fault affecting the heat transfer coefficient
0	initial conditions
$\star$	nominal value
$\hat{}$	estimate
$\tilde{}$	estimation error

6.1 Fault Diagnosis Strategies for Batch Reactors

In chemical processes, several kinds of failures may compromise safety and productivity. Indeed, the occurrence of faults may affect efficiency of the process (e.g., lower product quality) or, in the worst scenarios, could lead to fatal accidents (e.g., temperature runaway) with injuries to personnel, environmental pollution, equipments damage.

The term *fault* is generally defined as a departure of an observed variable or a parameter from an acceptable range [19, 48]. The causes of this abnormality, such as a failed coolant pump or a failed sensor, are called basic events or root events and are often referred as malfunctions or failures.

Fault Diagnosis (FD) consists of three main tasks:

- *fault detection*, i.e., the detection of the occurrence of a fault
- *fault isolation*, i.e., the determination of the type and/or location of the fault; and
- *fault identification*, i.e., the determination of the time evolution of the fault.

Once a fault has been detected, in some applications a controller reconfiguration for the self-correction of the fault is required (*fault accommodation*). A fault-tolerant control system possesses this capability, and its adoption is of the utmost importance when the controlled system can have potentially damaging effects on the environment if its components fail.

The relative importance of the three above-mentioned diagnosis tasks depends on the application at hand. However, while fault detection is an absolute must for any practical system, fault identification, even if helpful, is not essential when controller reconfiguration is not required.

Essentially, an FD system must avoid two kinds of errors, false alarms and missed alarms. A false alarm occurs when a fault is declared but the system is operating in healthy conditions; typically, they are due to model uncertainties and disturbances. On the other hand, a missed alarm occurs when, under faulty condition, the FD system does not detect any fault. Usually, minimization of false alarm and missed alarms are conflicting requirements.

Early approaches to fault diagnosis were often based on the so-called *physical redundancy* [11], i.e., the duplication of sensors, actuators, computers, and softwares to measure and/or control a variable. Typically, a voting scheme is applied to the redundant system to detect and isolate a fault. The physical redundant methods are very reliable, but they need extra equipment and extra maintenance costs. Thus, in the last years, researchers focused their attention on techniques not requiring extra equipment. These techniques can be classified into two general categories, model-free data-driven approaches and model-based approaches.

6.1.1 Model-Free Approaches

Model-free data-driven approaches do not require a model of the monitored process, but only a good database of historical data collected in normal operating conditions. This class of approaches includes both statistical and knowledge-based methods [49].

As concerns the former, statistical tests on the measured data are usually adopted to detect any abnormal behavior. In other words, an industrial process is considered as a stochastic system and the measured data are considered as different realizations of the stochastic process. The distribution of the observations in normal operating conditions is different from those related to the faulty process. Early statistical approaches are based on univariate statistical techniques, i.e., the distribution of a monitored variable is taken into account. For instance, if the monitored variable follows a normal distribution, the parameters of interest are the mean and standard deviation that, in faulty conditions, may deviate from their nominal values. Therefore, fault diagnosis can be reformulated as the problem of detecting changes in the parameters of a stochastic variable [3, 30].

However, since failures may involve a large number of parameters, often not independent from each other, the univariate techniques may be not so efficient; therefore, they have been replaced by multivariate techniques, which are powerful tools able to compress data and reduce the problem dimensionality while retaining the essential information. In detail, Principal Component Analysis (PCA) [12, 47] is a standard multivariate technique, whose main goal is to transform a number of related process variables to a smaller set of uncorrelated variables; Projection to Latent Structures (PLS) [29, 54] is conceptually similar to PCA but allows one to reduce the dimensions of both process variables and product quality variables to be analyzed. Finally, it is worth mentioning set-based approaches [26].

Knowledge-based expert system approaches require acquisition of knowledge of the process, a suitable choice of the knowledge representation, a careful encoding of the knowledge, and the development of inference procedures for diagnostic purposes. The main drawback of expert systems is the unpredictability of the response of the system outside the domain of its expertise. This problem can be overcome by using Artificial Neural Networks (ANNs) that do not require explicit encoding of knowledge or an accurate knowledge of the mathematical model of the process and are characterized by good generalization capabilities. Early approaches include the use of back-propagation networks [20, 25, 48]. Then, a number of more efficient network structures have been considered: for example, Radial Basis Functions (RBFs) [55] have been suited to fault diagnosis for their better generalization performance. More recently, Bayesian belief networks [32, 33, 40], dynamic neural networks [34], and pattern recognition methods [31] have been proposed. In [41, 42], an approach based on a combination of ANNs and knowledge-based expert systems has been developed.

6.1.2 Model-Based Approaches

Model-based approaches to fault diagnosis can be divided into qualitative methods [51] and quantitative methods [35, 36].

Among qualitative methods, the most successful for chemical processes are those based on causal models, such as digraphs or fault-trees. Signed Digraphs (SDGs), i.e., graphs with directed arcs between the nodes, are built in such a way that nodes correspond to events, while edges represent the causal relationships between the events. SDGs have been the most widely used form of causal knowledge for process fault diagnosis until the late 1990s [6, 43, 46].

Fault trees are logic trees that propagate primary events or faults to the top level events. They are widely used for analyzing fault effects, as well as system reliability and safety. According to [15], fault tree analysis includes four steps: system definition, fault tree construction, qualitative evaluation, and quantitative evaluation. Each node of the tree represents a logical operation (AND or OR). Starting from the top event (i.e., the fault), the lower nodes are introduced by considering the events causing the top event, until events which cannot be further expanded are encountered. A major drawback of fault trees is that formal methods to verify the accuracy of the developed fault tree are still not available.

In the last two decades, the researchers' interest has been focused mainly on quantitative model-based methods, based on the concept of *analytical* or *functional redundancy*, which use a mathematical model of the process to obtain the estimates of a set of variables characterizing the behavior of the monitored system. The inconsistencies between estimated and measured variables provide a set of *residuals*, sensitive to the occurrence of faults. Later, the residuals are evaluated in order to identify and localize faults.

Although there is a close relationship among the various quantitative model-based techniques, observer-based approaches have become very important and diffused, especially within the automatic control community. Luenberger observers [1, 45, 53], unknown input observers [44], and Extended Kalman Filters [21] have been mostly used in fault detection and identification for chemical processes and plants. Reviews of several model-based techniques for FD can be found in [8, 13, 35, 50] and, as for the observer-based methods, in [1, 36, 44].

The literature focused on model-based FD presents a few applications of observers to chemical plants. In [10] an unknown input observer is adopted for a CSTR, while in [7] and [21] an Extended Kalman Filter is used; in [9] and [28] Extended Kalman Filters are used for a distillation column and a CSTR, respectively; in [45] a generalized Luenberger observer is presented; in [24] a geometric approach for a class of nonlinear systems is presented and applied to a polymerization process; in [38] a robust observer is used for sensor faults detection and isolation in chemical batch reactors, while in [37] the robust approach is compared with an adaptive observer for actuator fault diagnosis.

Since perfect knowledge of the model is rarely a reasonable assumption, soft computing methods, integrating quantitative and qualitative modeling information, have been developed to improve the performance of observer-based schemes for uncertain systems [36]. Major contributions to observer-based approaches can be found in [39, 56] as well, where fault isolation is achieved via a bank of observers, while identification is based on the adoption of online universal interpolators (e.g., ANNs whose weights are updated on line). As for the use of observers in the presence of advanced control techniques, such as MPC or FLC, in [44] an unknown input observer is adopted in conjunction with an MPC scheme.

However, most of the above-mentioned approaches are referred to continuous reactors; application of these techniques to batch chemical processes is usually difficult, because of their nonlinear dynamics, intrinsically unsteady operating conditions, lack of full state measurements, and poor model knowledge.

In this chapter, an FD framework for batch chemical processes is developed, where diagnosis of sensor, actuator, and process faults can be achieved via an integrated approach. The proposed approach is based on physical redundancy for detection of sensor faults [38], while an analytical redundancy method, based on a bank of diagnostic observers, is adopted to perform process/actuator fault detection, isolation, and identification [4].

6.2 Basic Principles of Model-Based Fault Diagnosis

Consider a nonlinear time-varying dynamic system

$$\begin{cases} \dot{x} = g(t, x, u) + \eta(t, x, u), \\ y = h(t, x, u), \end{cases} \tag{6.1}$$

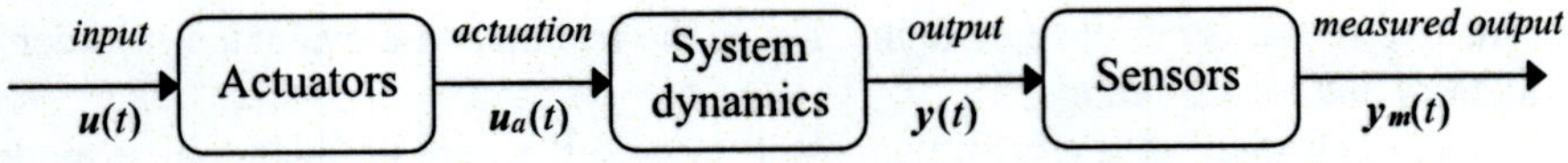

Fig. 6.1 System modeling

where, as usual, $\boldsymbol{x}$, $\boldsymbol{u}$, and $\boldsymbol{y}$ are the state, input, and output vectors, respectively, while $\boldsymbol{\eta}$ represents the system uncertainties vector.

The information used for FD are the measured output from sensor and the input to the actuators. Therefore, for FD purposes, the open-loop system is usually considered as composed by three main blocks [8]: actuators, process, and sensors. Namely, in Fig. 6.1, $\boldsymbol{u}$ is the system input (control action), $\boldsymbol{u}_{\mathrm{a}}$ is the actuator output, $\boldsymbol{y}$ is the system output, and $\boldsymbol{y}_{\mathrm{m}}$ is the output measured by sensors.

If the dynamics of sensors and actuators is neglected, the fault-free condition is characterized by the following relations:

$$\boldsymbol{y}_{\mathrm{m}}(t) = \boldsymbol{y}(t) + \boldsymbol{n}(t), \qquad \boldsymbol{u}_{\mathrm{a}}(t) = \boldsymbol{u}(t),$$

where $\boldsymbol{n}$ is the vector of measurement noise. A sensor fault can be described mathematically as

$$\boldsymbol{y}_{\mathrm{m}}(t) = \boldsymbol{y}(t) + \boldsymbol{n}(t) + \boldsymbol{f}_{\mathrm{s}}(t), \tag{6.2}$$

where $\boldsymbol{f}_{\mathrm{s}}$ is the sensor fault vector. The above model, although very simple, may capture all main sensor fault classes, provided that $\boldsymbol{f}_{\mathrm{s}}$ is suitably chosen. For instance, an abrupt switch to zero of the measured signal is described by $\boldsymbol{f}_{\mathrm{s}}(t) = -\boldsymbol{y}(t)$, while an abrupt constant bias $\delta\boldsymbol{y}$, affecting the measured signal, can be modeled as $\boldsymbol{f}_{\mathrm{s}}(t) = \delta\boldsymbol{y}$.

In a similar way, the actuator action, in the presence of an actuator fault, becomes

$$\boldsymbol{u}_{\mathrm{a}}(t) = \boldsymbol{u}(t) + \boldsymbol{f}_{\mathrm{a}}(t), \tag{6.3}$$

where $\boldsymbol{f}_{\mathrm{a}}$ is the actuator fault vector. A wide class of actuator faults can be represented by (6.3), provided that $\boldsymbol{f}_{\mathrm{a}}$ is suitably chosen. For instance, an abrupt constant bias affecting the actuator output, $\delta\boldsymbol{u}$, can be modeled as $\boldsymbol{f}_{\mathrm{a}}(t) = \delta\boldsymbol{u}$; if the actuator output is frozen at the value $\boldsymbol{u}(t^*)$ for $t \geq t^*$, the corresponding fault vector is $\boldsymbol{f}_{\mathrm{a}}(t) = \boldsymbol{u}(t^*) - \boldsymbol{u}(t)$.

As for process faults, they can be modeled via an unknown additional term $\boldsymbol{f}_{\mathrm{p}}(t, \boldsymbol{x}, \boldsymbol{u})$ affecting the state equation of the system dynamics (6.1).

As previously stated, model-based FD consists on detection, isolation, and identification of faults in the components of a system from the comparison of the system measurements with a priori information given by the mathematical model. The discrepancies between the real system behavior and the behavior predicted by the model are taken into account via quantities called *residuals*. Then, the residuals are processed by a decision making system (Fig. 6.2) whose aim is to generate alarms and/or directions for other subsystems (e.g., the control system).

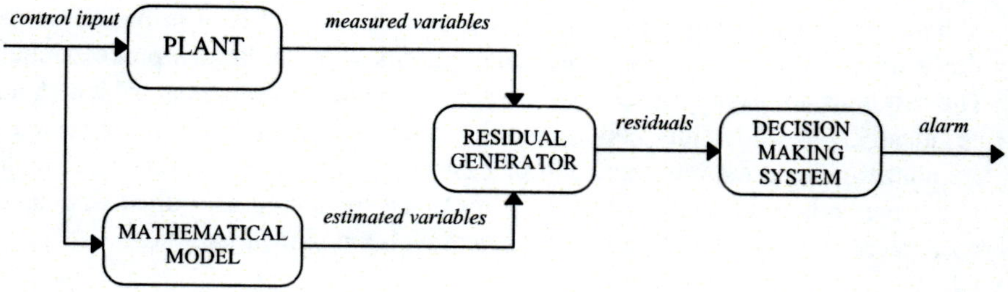

Fig. 6.2 Scheme of a model-based diagnosis system

6.2.1 Residual Generation

In general, residuals can be obtained as functions of the measured output $\boldsymbol{y}_{\mathrm{m}}$ and its estimate $\widehat{\boldsymbol{y}}$ computed via the available mathematical model

$$\boldsymbol{r} = \boldsymbol{r}(t, \boldsymbol{y}_{\mathrm{m}}, \widehat{\boldsymbol{y}}). \tag{6.4}$$

Ideally, residuals should be equal to zero in the absence of faults, while they should become nonzero after the occurrence of faults. Of course, in practice, they are always nonzero due to model uncertainties and disturbances. Since the residual generation is the most important issue of quantitative model-based fault diagnosis, most of the works in this research field have been focused on this problem. A wide variety of techniques are available in the literature (see, e.g., [8, 16] for a complete overview). Since a complete review is outwith the scope of this book, in the following, only the basic concepts of the main approaches are briefly discussed.

The basic idea of observer-based approaches is to estimate the outputs of the system from available measurements by using suitably defined *diagnostic observers*. Then, the output estimation errors are used as residuals. For isolation purposes, a bank of observers can be adopted, where each observer computes an estimate of the system state sensitive to all faults but one.

Approaches based on parameter estimation assume that the faults lead to detectable changes of physical system parameters. Therefore, FD can be pursued by comparing the estimates of the system parameters with the nominal values obtained in healthy conditions. The operative procedure, originally established in [23], requires an accurate model of the process (including a reliable nominal estimate of the model parameters) and the determination of the relationship between model parameters and physical parameters. Then, an online estimation of the process parameters is performed on the basis of available measures. This approach, of course, might reveal ineffective when the parameter estimation technique requires solution to a nonlinear optimization problem. In such cases, reduced-order or simplified mathematical models may be used, at the expense of accuracy and robustness. Moreover, fault isolation could be difficult to achieve, since model parameters cannot always be converted back into corresponding physical parameters, and thus the influence of each physical parameters on the residuals could not be easily determined.

The basic idea of the parity equation approach [18, 52] is to provide a proper check of the consistency of the measured outputs with the known process inputs. The residuals are usually given by the value of the parity equations, which should be, ideally, zero in healthy conditions. Of course, in real situations, the residuals are nonzero due to measurement and process noise, model inaccuracies, and faults. These methods usually adopt linear or linearized models; thus, difficulties can be encountered when dealing with complex and highly nonlinear systems.

6.2.2 Decision Making System and Fault Isolation

After their generation, residuals must be processed in order to detect and isolate faults. The decision process usually comes down to the selection of suitable thresholds. In ideal conditions (i.e., residuals not affected by uncertainties and disturbances), thresholds can be set to zero. However, in practice, this condition cannot be achieved; therefore, nonzero thresholds must be adopted.

The simplest strategy is to set fixed thresholds, chosen in such a way to find a compromise between sensitivity to faults and the need to minimize false alarms. In fact, setting too low thresholds might increase the rate of false alarms, while choosing them too large might reduce sensitivity. As for the thresholds selection, different approaches may be pursued. A priori selection of the thresholds, i.e., obtained by computing the upper bounds of the residuals in the absence of faults, may result in too conservative and thus compromising sensitivity to faults. In alternative, an empirical approach may be adopted. Namely, a number of experiments, in the absence of faults, can be performed, and the corresponding residuals recorded; then, the thresholds can be set on the basis of the values of the residuals in healthy conditions.

Sensitivity to faults can be improved by using adaptive thresholds, adjusted online on the basis of measurements [56]. Adaptive thresholds may help in evaluating residuals, provided that suitable bounding functions for the uncertainties are devised. In the case of complex systems, the sources of uncertainties are often difficult to model; hence, it can be quite challenging to derive reliable bounding functions.

A different approach to threshold selection is based on fuzzy logic [14]. By adopting this approach, the value that represents the crisp discriminant between faults and disturbances is replaced by a fuzzy set, characterized by a membership function. Hence, a yes–no decision is replaced by a continuous indication of the faulty level.

Whilst a single residual may be sufficient to detect faults, a vector of residuals is usually required for fault isolation. For isolation purposes, *structured residuals* [8, 17] can be generated, i.e., each residual is affected only by a specific subset of faults, and each fault only affects a specific subset of residuals. This concept can be expressed in a mathematical form by introducing a boolean fault code vector and a boolean structure matrix [8].

In this chapter, a bank of observers is adopted for isolation of process and actuator faults. Namely, it is assumed that only N_{F} different types of faults can occur. Then, each isolation observer is designed so that its output is sensitive to all faults in the set but one. In this way, a set of isolation residuals can be computed, where each of them is sensitive to all faults but one.

6.3 Fault Diagnosis for Chemical Batch Reactors

Consider a jacketed batch reactor in which the same network of irreversible exothermic reactions adopted in Chap. 5 takes place. In order to design the FD framework, the state-space model (5.15) is rewritten in a slightly different form, i.e.,

$$\begin{cases} \dot{\boldsymbol{x}} = \boldsymbol{A}_{\mathrm{d}}(\boldsymbol{y})\boldsymbol{x} + \boldsymbol{b}(\boldsymbol{y}, u) + \boldsymbol{\eta}(t, \boldsymbol{x}, u), \\ \boldsymbol{y} = \boldsymbol{C}\boldsymbol{x} + \boldsymbol{n}, \end{cases} \tag{6.5}$$

where the state vector $\boldsymbol{x}$, the input u, the output vector $\boldsymbol{y}$, the vector $\boldsymbol{b}$, and the matrix $\boldsymbol{C}$ are the same defined in Sect. 5.5, while the matrix $\boldsymbol{A}_{\mathrm{d}}(\boldsymbol{y})$ is given by

$$\boldsymbol{A}_{\mathrm{d}} = \begin{bmatrix} \boldsymbol{A}_{\mathrm{M}}(\boldsymbol{y}) & \boldsymbol{O}_{N_{\mathrm{C}} \times 2} \\ \boldsymbol{A}_{\mathrm{M,E}}(\boldsymbol{y}) & \boldsymbol{A}_{\mathrm{E}} \end{bmatrix}, \tag{6.6}$$

where $\boldsymbol{A}_{\mathrm{M}}$, $\boldsymbol{A}_{\mathrm{M,E}}$, and $\boldsymbol{A}_{\mathrm{E}}$ are given in Sects. 5.5 and 5.6.2.1. The matrix $\boldsymbol{A}_{\mathrm{E}}$ contains the available estimate $\theta^{\star}$ of the parameter θ, since all model uncertainties (e.g., parametric uncertainties, effects of nonideal mixing and/or heating/cooling, heat losses) have been lumped into the $((N_{\mathrm{C}} + 2) \times 1)$ vector $\boldsymbol{\eta}$. Finally, the (2×1) vector $\boldsymbol{n}$ represents the measurement noise.

Since a fairly accurate model of the batch reactor can be derived according to the guidelines given in Chaps. 2 and 3, the existence of a finite (and relatively small) bound on the magnitude of uncertainties can be assumed. Moreover, it is customary to assume bounded sensor noise as well. Hence, the following assumption is made:

Assumption 6.1 The vectors $\boldsymbol{\eta}$ and $\boldsymbol{n}$ are norm-bounded, i.e.,

$$\exists \bar{\eta} > 0: \quad \left\| \boldsymbol{\eta}(t, \boldsymbol{x}, u) \right\| \leq \bar{\eta} \quad \forall t \in \mathcal{T}, \boldsymbol{x} \in \mathcal{X}, \ \forall u \in \mathcal{U},$$

$$\exists \bar{n} > 0: \quad \left\| \boldsymbol{n}(t) \right\| \leq \bar{n} \quad \forall t \in \mathcal{T},$$

where $\mathcal{X}$ and $\mathcal{U}$ are the admissible sets to which the state and the input belong, respectively, while $\mathcal{T}$ is the considered time interval.

6.3.1 Fault Characterization

Faults can occur due to sensors failures, equipment failures, or changes in process parameters. As previously stated in Sect. 6.2, sensor faults can be modeled as an unknown additive term in the output equation, i.e.,

$$y(t) = Cx(t) + f_s(t) + n(t), \tag{6.7}$$

where the (2×1) vector f_s collects the unknown faults profiles affecting the vessel and the jacket temperature measurements, respectively. They could be due to, e.g., a constant bias superimposed on the measured variable or a drift of the measured value.

An actuator fault can be generated by a malfunction of the cooling system, such as electric-power failures, pomp failures, valves failures, and leaking pipes. Without loss of generality, actuator faults may be modeled as an unknown additive term affecting the state equation in (6.5), due to unexpected variations of the input u with respect to its nominal value, i.e., the value computed by the reactor control system.

Process faults can be seen as environmental disturbances affecting the process. Some examples for this kind of faults are effects of side reactions (caused, e.g., by impurities in raw materials) or unexpected changes in the heat transfer coefficient (due, e.g., to fouling in the heat exchanger). In this chapter, only process faults affecting the dynamics of the state variables x_E are taken into account and are modeled as an additive term affecting the first equation in (6.5). This choice is due to the fact that the effect of process faults affecting the dynamics of x_M cannot be directly observed, since the measurements of reactants concentrations are assumed to be not available; however, this choice is only apparently limiting, since a wide class of faults affecting the dynamics of x_M (e.g., effects of side reactions) can be indirectly detected by modeling their effect on the reactor temperature.

In sum, the effects of both process and actuator faults on the system dynamics can be modeled via an additive term $C^T f_a(t, y, u)$ in the state equation in (6.5). As customary in the literature (see, e.g., [56]), the function f_a is assumed to belong to a finite set of N_F functions

$$\mathcal{F}_a = \{f_{a,1}, \dots, f_{a,N_F}\}.$$

Each fault function in $\mathcal{F}_a$ is assumed to have a linear-in-the-parameters structure, i.e.,

$$f_{a,i}(t, y, u) = \varphi_i(t, y, u)\theta_{f,i}, \quad i = 1, \dots, N_F, \tag{6.8}$$

where φ_i is a known $(2 \times m_i)$ regressor matrix, while $\theta_{f,i}$ is an unknown $(m_i \times 1)$ vector of constant parameters. The following standard assumption is adopted:

Assumption 6.2 The regressor matrix φ_i is assumed norm-bounded for all fault types, i.e., for all $i = 1, \dots, N_F$,

$$\exists \bar{\varphi}_i > 0: \quad \left\| \varphi_i(t, y, u) \right\| \le \bar{\varphi}_i \quad \forall t \in \mathcal{T},\ \forall y \in \mathcal{Y},\ \forall u \in \mathcal{U}.$$

It is worth noticing that a general class of faults can be effectively modeled via (6.8), where φ_i takes into account the structure of the fault, while $\theta_{f,i}$ characterizes its magnitude. Nevertheless, when the faults are not characterized by a linear-in-the-parameters structures, some well-established approaches in the literature adopt the so-called online interpolators [56] (e.g., neural networks, splines) to obtain a good approximation of the fault via parametric models.

Therefore, in the presence of faults, the state-space model (6.5) becomes

$$\begin{cases} \dot{\boldsymbol{x}} = \boldsymbol{A}_{\mathrm{d}}(\boldsymbol{y})\boldsymbol{x} + \boldsymbol{b}(\boldsymbol{y}, u) + \boldsymbol{C}^{\mathrm{T}} \boldsymbol{f}_{\mathrm{a}}(t, \boldsymbol{y}, u) + \boldsymbol{\eta}(t, \boldsymbol{x}, u), \\ \boldsymbol{y} = \boldsymbol{C}\boldsymbol{x} + \boldsymbol{f}_{\mathrm{s}} + \boldsymbol{n}, \end{cases} \tag{6.9}$$

where it is assumed that $\boldsymbol{f}_{\mathrm{s}}$ and $\boldsymbol{f}_{\mathrm{a}}$ are null before the occurrence of a fault at $t = t_{\mathrm{f}}$, i.e., $\boldsymbol{f}_{\mathrm{s}}(t) = \boldsymbol{0}$ and $\boldsymbol{f}_{\mathrm{a}}(t, \boldsymbol{y}(t), u(t)) = \boldsymbol{0}$ for $t < t_{\mathrm{f}}$.

The above model includes the case in which a sensor and a process/actuator fault occur during the same batch operation. However, occurrence of multiple faults of the same nature (i.e., multiple process/actuator faults or multiple sensor faults), is not considered.

Assumption 6.3 It is assumed that multiple process/actuator faults (i.e., two or more faults belonging to $\mathcal{F}_{\mathrm{a}}$ occur) and multiple sensor faults (i.e., two or more sensors are subject to failures) cannot occur during the same batch operation.

In the following, it is shown that multiple process/actuator faults (multiple sensor faults) can be detected but not correctly isolated and identified.

On the contrary, occurrence of sensor and process/actuator faults during the same batch is allowed.

6.3.2 Architecture of the Fault Diagnosis Scheme

Due to the level of risk related to highly exothermic chemical processes, sensors for temperature monitoring are often duplicated in batch reactors. Hence, a duplex sensor architecture is assumed. Namely, two temperature sensors (hereafter labeled as $S_{\mathrm{r},1}$ and $S_{\mathrm{r},2}$) providing measurements of T_{r}, and two providing measurements of T_{j} (hereafter labeled as $S_{\mathrm{j},1}$ and $S_{\mathrm{j},2}$) are assumed to be available.

In order to achieve both sensor fault detection and isolation, two state observers of the system are adopted: the first observer uses the measurements provided by $S_{\mathrm{r},1}$ and $S_{\mathrm{j},1}$, while the second observer uses the measurements provided by $S_{\mathrm{r},2}$ and $S_{\mathrm{j},2}$.

A suitable designed diagnostic system, together with a Decision Making System (DMS), declares the occurrence of a fault, isolates the possible faulty sensor, and outputs an healthy signal.

Then, the healthy signal is used to feed a bank of $N_{\mathrm{F}} + 1$ nonlinear adaptive observers (where N_{F} is the number of the possible process/actuator faults). The first observer is in charge of detecting the occurrence of process/actuator faults. The other N_{F} observers, each corresponding to a particular type of process/actuator fault, achieve fault isolation and identification by adopting a suitable adaption mechanism. Figure 6.3 shows a block diagram representation of the overall architecture.

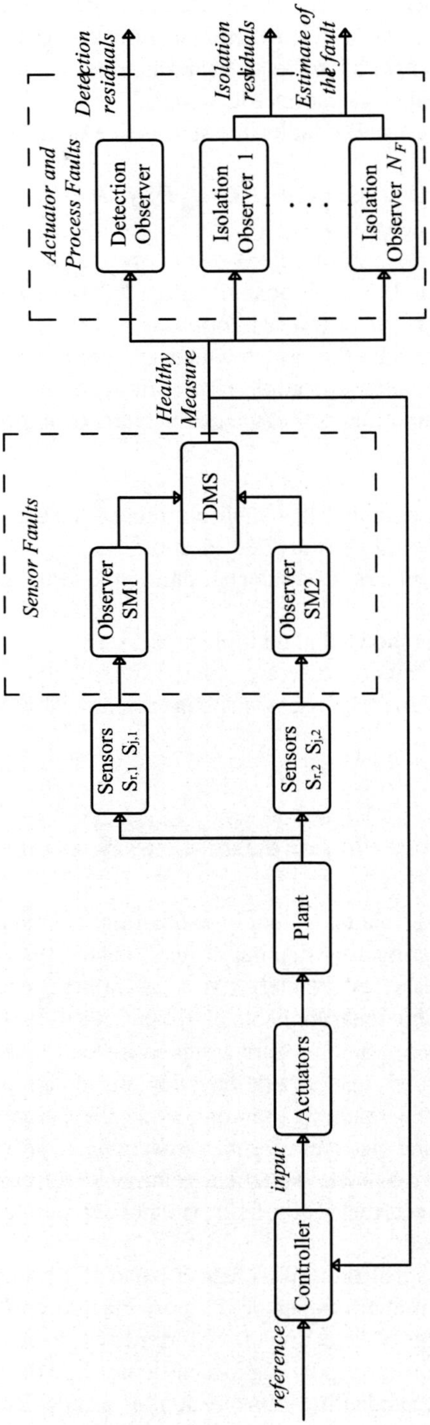

Fig. 6.3 Architecture of the fault detection and isolation scheme

6.4 Sensor Fault Diagnosis

As previously stated, two observers are adopted for sensor fault diagnosis:

- observer SM1 uses the measurements provided by $S_{r,1}$ and $S_{j,1}$, i.e., the output vector $\boldsymbol{y}_{SM1} = (y_{r,1}\ y_{j,1})^T$; and
- observer SM2 uses the measurements provided by $S_{r,2}$ and $S_{j,2}$, i.e., the output vector $\boldsymbol{y}_{SM2} = (y_{r,2}\ y_{j,2})^T$.

Both the observers have the following form (hereafter $i = 1, 2$):

$$\begin{cases} \dot{\widehat{\boldsymbol{x}}}_{SMi} = \boldsymbol{A}_d(\boldsymbol{y}_{SMi})\widehat{\boldsymbol{x}}_{SMi} + \boldsymbol{b}(\boldsymbol{y}_{SMi}, u) + \boldsymbol{L}_s\widetilde{\boldsymbol{y}}_{SMi}, \\ \widehat{\boldsymbol{y}}_{SMi} = \boldsymbol{C}\widehat{\boldsymbol{x}}_{SMi}, \end{cases} \tag{6.10}$$

where $\widehat{\boldsymbol{x}}_{SMi}$ denotes the vector of the state estimates, $\widehat{\boldsymbol{y}}_{SMi}$ and $\widetilde{\boldsymbol{y}}_{SMi} = \boldsymbol{y}_{SMi} - \widehat{\boldsymbol{y}}_{SMi}$ denote the vectors of output estimates and output estimation errors, respectively, and $\boldsymbol{L}_s$ is an $(N_C + 2) \times 2$ matrix of gains defined as follows:

$$\boldsymbol{L}_s = \begin{bmatrix} \boldsymbol{L}_M \\ \boldsymbol{L}_E \end{bmatrix}, \qquad \boldsymbol{L}_M = \begin{bmatrix} l_1 & 0 \\ l_2 & 0 \\ \vdots & \vdots \\ l_{N_C} & 0 \end{bmatrix}, \qquad \boldsymbol{L}_E = \begin{bmatrix} l_r & 0 \\ 0 & l_j \end{bmatrix}.$$

The state estimation error $\widetilde{\boldsymbol{x}}_{SMi} = \boldsymbol{x} - \widehat{\boldsymbol{x}}_{SMi}$ can be analyzed by considering the estimation error dynamics, derived from (6.5) and (6.10),

$$\begin{cases} \dot{\widetilde{\boldsymbol{x}}}_{SMi} = \boldsymbol{A}_{s,i}(\boldsymbol{y}_{SMi})\widetilde{\boldsymbol{x}}_{SMi} + \boldsymbol{\eta}_{s,i}(t, \boldsymbol{x}, u) + \boldsymbol{L}_s\boldsymbol{f}_s(t), \\ \widetilde{\boldsymbol{y}}_{SMi} = \boldsymbol{C}\widetilde{\boldsymbol{x}}_{SMi} + \boldsymbol{f}_s(t) + \boldsymbol{n}_i, \end{cases} \tag{6.11}$$

where $\boldsymbol{A}_{s,i}(\boldsymbol{y}_{SMi}) = \boldsymbol{A}_d(\boldsymbol{y}_{SMi}) - \boldsymbol{L}_s\boldsymbol{C}$, $\boldsymbol{\eta}_{s,i} = \boldsymbol{\eta} + \boldsymbol{L}_s\boldsymbol{n}_i$, and $\boldsymbol{n}_i$ represents the measurement noise affecting the ith couple of sensors (upper bounded by $\bar{n}$, as stated by Assumption 6.1).

Convergence properties of $\widetilde{\boldsymbol{x}}_{SMi}$ are established by the following result.

Theorem 6.1 *In the absence of faults, uncertainties, and sensor noise (i.e.,* $\boldsymbol{f}_s = \boldsymbol{0}$, $\boldsymbol{\eta} = \boldsymbol{0}$, *and* $\boldsymbol{n}_i = \boldsymbol{0}$*), if the rate constants are bounded as in* (2.32) *and* (2.33)*, there exists a set of observer gains such that the state estimation error* $\widetilde{\boldsymbol{x}}_{SMi}$ *of the observer* (6.10) *is globally uniformly convergent to* $\boldsymbol{0}$ *as* $t \to \infty$*. Moreover, the convergence is exponential.*

The above result can be proven by using the same arguments used in the proof of Theorem 5.1; for the sake of completeness, the proof is reported in Appendix A.5.

Then, the behavior of the estimation error in the absence of faults ($\boldsymbol{f}_s = \boldsymbol{0}$) and in the presence of bounded uncertainties and sensor noise is considered. In view of (6.11), the evolution of $\widetilde{\boldsymbol{x}}_{SMi}$ starting from the initial time t_0, given the initial state estimation error $\widetilde{\boldsymbol{x}}_{\mathrm{SM}i}(t_0)$, can be expressed as follows:

$$\widetilde{\boldsymbol{x}}_{\mathrm{SM}i}(t) = \boldsymbol{\Phi}_{\mathrm{s},i}(t,t_0)\widetilde{\boldsymbol{x}}_{\mathrm{SM}i}(t_0) + \int_{t_0}^{t} \boldsymbol{\Phi}_{\mathrm{s},i}(t,\zeta)\boldsymbol{\eta}_{\mathrm{s},i}\big(\zeta,\boldsymbol{x}(\zeta),\boldsymbol{u}(\zeta)\big)\,d\zeta \quad \forall t \geq t_0, \tag{6.12}$$

where $\boldsymbol{\Phi}_{\mathrm{s},i}$ denotes the state transition matrix [27] corresponding to $\boldsymbol{A}_{\mathrm{s},i}$. Thanks to the global asymptotic convergence of $\widetilde{\boldsymbol{x}}_{\mathrm{SM}i}$, stated by Theorem 6.1, the following inequality holds [27]:

$$\exists \kappa_{\mathrm{s},i} > 0, \lambda_{\mathrm{s},i} > 0: \quad \big\|\boldsymbol{\Phi}_{\mathrm{s},i}(t,t_0)\big\| \leq \kappa_{\mathrm{s},i} e^{-\lambda_{\mathrm{s},i}(t-t_0)} \quad \forall t \geq t_0. \tag{6.13}$$

Hence, in view of (6.13) and Assumption 6.2, in the absence of faults, the norm of the output estimation error can be upper bounded as follows:

$$\begin{aligned}
\big\|\widetilde{\boldsymbol{y}}_{\mathrm{SM}i}(t)\big\| &= \big\|\boldsymbol{C}\widetilde{\boldsymbol{x}}_{\mathrm{SM}i}(t) + \boldsymbol{n}_i(t)\big\| \leq \big\|\widetilde{\boldsymbol{x}}_{\mathrm{SM}i}(t)\big\| + \big\|\boldsymbol{n}_i(t)\big\| \\
&\leq \left\| \boldsymbol{\Phi}_{\mathrm{s},i}(t,t_0)\widetilde{\boldsymbol{x}}_{\mathrm{SM}i}(t_0) + \int_{t_0}^{t} \boldsymbol{\Phi}_{\mathrm{s},i}(t,\zeta)\boldsymbol{\eta}_{\mathrm{s},i}\big(\zeta,\boldsymbol{x}(\zeta),\boldsymbol{u}(\zeta)\big)\,d\zeta \right\| + \big\|\boldsymbol{n}_i(t)\big\| \\
&\leq \kappa_{\mathrm{s},i} e^{-\lambda_{\mathrm{s},i}(t-t_0)} \big\|\widetilde{\boldsymbol{x}}_{\mathrm{SM}i}(t_0)\big\| + \int_{t_0}^{t} \kappa_{\mathrm{s},i} e^{-\lambda_{\mathrm{s},i}(t-\zeta)} \bar{\eta}_{\mathrm{s}}\,d\zeta + \bar{n} \\
&= \kappa_{\mathrm{s},i} e^{-\lambda_{\mathrm{s},i}(t-t_0)} \big\|\widetilde{\boldsymbol{x}}_{\mathrm{SM}i}(t_0)\big\| + \frac{\kappa_{\mathrm{s},i}\bar{\eta}_{\mathrm{s}}}{\lambda_{\mathrm{s},i}}\big(1 - e^{-\lambda_{\mathrm{s},i}(t-t_0)}\big) + \bar{n} \\
&\leq \kappa_{\mathrm{s},i}\left(\big\|\widetilde{\boldsymbol{x}}_{\mathrm{SM}i}(t_0)\big\| + \frac{\bar{\eta}_{\mathrm{s}}}{\lambda_{\mathrm{s},i}} \right) + \bar{n} = \bar{\mu}_{\mathrm{s},i} \quad \forall t \geq t_0,
\end{aligned} \tag{6.14}$$

where $\bar{\eta}_{\mathrm{s}} = \bar{\eta} + \|\boldsymbol{L}_{\mathrm{s}}\|\bar{n}$.

In sum, thanks to the convergence properties stated by Theorem 6.1, the output estimation error of each observer keeps bounded in the presence of bounded uncertainties and sensor noise. It is worth noticing that the bound $\bar{\mu}_{\mathrm{s},i}$ could be, in principle, determined if all the constants needed for its computation (i.e., $\bar{\eta}_{\mathrm{s}}$, $\bar{n}$, $\kappa_{\mathrm{s},i}$, $\lambda_{\mathrm{s},i}$, and $\widetilde{\boldsymbol{x}}_{\mathrm{SM}i}(t_0)$) are known or, at least, estimated with reasonable accuracy. Nevertheless, in practice, such a bound may be quite conservative and thus useless.

Remark 6.1 It can be shown [5] that the bound on $\|\widetilde{\boldsymbol{y}}_{\mathrm{SM}i}(t)\|$ decreases with the largest eigenvalue of the matrix $\boldsymbol{A}_{\mathrm{E}} - \boldsymbol{L}_{\mathrm{E}}$, λ_{E}, and the norm of the initial state estimation error; thus, it can be reduced if a suitably large value for λ_{E} is chosen and a good initial guess of the state is available. As for the convergence rate of $\widetilde{\boldsymbol{y}}_{\mathrm{SM}i}$, in [5] it is shown that, for suitably large values of λ_{E}, the dynamics of $\widetilde{\boldsymbol{y}}_{\mathrm{SM}i}$ is mainly determined by λ_{E} itself and not by the (usually slower) dynamics of the reactive process.

Table 6.1 Decisions made by the diagnostic system on the basis of the residuals

$\lvert r_{S_r}\rvert > 1$	$\lvert r_{S_j}\rvert > 1$	$\lVert \boldsymbol{r}_{SM1}\rVert > 1$	$\lVert \boldsymbol{r}_{SM2}\rVert > 1$	Decision
No	No	No	No	Fault not declared
Yes	No	Yes	No	Fault declared on sensor $S_{r,1}$
Yes	No	No	Yes	Fault declared on sensor $S_{r,2}$
Yes	No	No	No	Fault declared but not isolated
No	Yes	Yes	No	Fault declared on sensor $S_{j,1}$
No	Yes	No	Yes	Fault declared on sensor $S_{j,2}$
No	Yes	No	No	Fault declared but not isolated

6.4.1 Residuals Generation and Fault Isolation

Detection of sensor faults can be achieved on the basis of the following residuals:

$$r_{S_r} = \frac{y_{r,1} - y_{r,2}}{\mu_{s,r}}, \qquad r_{S_j} = \frac{y_{j,1} - y_{j,2}}{\mu_{s,j}}, \tag{6.15}$$

where $\mu_{s,r}$ and $\mu_{s,j}$ are normalization factors to be properly determined. Hence, if either $S_{r,1}$ or $S_{r,2}$ (either $S_{j,1}$ or $S_{j,2}$) is affected by a fault, the absolute value of r_{S_r} (r_{S_j}) is expected to exceed a certain threshold. The normalization factors can be selected so as to set the threshold to 1. Hence, a possible choice is given by $\mu_{s,r} = \mu_{s,j} = 2\bar{n}$, since, in the absence of faults, $|y_{*,1}(t) - y_{*,2}(t)| \le 2\bar{n}$ $(* = \text{r}, \text{j})$ for all t. Then, it can be easily shown that a sensor fault, occurring at $t = t_f$, can be detected if its absolute value exceeds the quantity $\mu_{s,*} + 2\bar{n}$ for some $t \ge t_f$.

For isolation purposes (i.e., determination of the faulty sensor), two other residuals must be defined,

$$\boldsymbol{r}_{SM1} = \frac{\widetilde{\boldsymbol{y}}_{SM1}}{\mu_{s,1}}, \qquad \boldsymbol{r}_{SM2} = \frac{\widetilde{\boldsymbol{y}}_{SM2}}{\mu_{s,2}}, \tag{6.16}$$

where $\mu_{s,1}$ and $\mu_{s,2}$ are normalization factors, which can be set, e.g., on the basis of experimental data collected in the absence of faults. According to (6.14), a possible choice for the normalization factors is the value $\bar{\mu}_{s,i}$. By virtue of these normalization factors, the thresholds on the residuals can be set to 1, and the norm of residual vectors can be used to isolate faults. In fact, the output of the SM1 observer is not affected by faults on $S_{r,2}$ and $S_{j,2}$, while the output of the SM2 observer is not affected by faults on $S_{r,1}$ and $S_{j,1}$. Hence, if the norm of $\boldsymbol{r}_{SM1}$ ($\boldsymbol{r}_{SM2}$) exceeds the threshold, a fault is declared on either $S_{r,1}$ or $S_{j,1}$ (either $S_{r,2}$ or $S_{j,2}$), depending on which detection residual (i.e., r_{S_r} or r_{S_j}) exceeds the threshold.

In sum, a fault can be declared and, eventually, isolated, provided that simultaneous faults on different sensors do not occur during the same batch operation, according to the logic summarized in Table 6.1.

When a sensor fault (occurring at $t = t_f$) affects one of the sensors in the couple $\{S_{r,i}, S_{j,i}\}$ $(i = 1, 2)$, the following equality holds:

$$\widetilde{y}_{SMi}(t) = C\widetilde{x}_{SMi}(t) + f_s(t) + n_i(t) \quad \forall t \geq t_f. \tag{6.17}$$

According to (6.14), the following inequality can be derived:

$$\|\widetilde{y}_{SMi}(t)\| \geq \|f_s(t)\| - \bar{\mu}_{s,i} \quad \forall t \geq t_f.$$

Therefore, a sufficient condition ensuring isolation of a fault affecting sensor $S_{*,i}$ $(* = r, j)$ is

$$\exists t > t_f: \quad |y_{*,1}(t) - y_{*,2}(t)| > \mu_{s,*} \quad \text{and} \quad \|f_s(t)\| > \bar{\mu}_{s,i} + \mu_{s,i}, \tag{6.18}$$

and (for $l \neq i$)

$$\|\widetilde{y}_{SMl}\| \leq \bar{\mu}_{s,l} \quad \forall t > t_f. \tag{6.19}$$

In fact, the above condition guarantees that $\|r_{SMi}\| = \|\widetilde{y}_{SMi}\|/\mu_{s,i}$ exceeds 1 at least for a time instant after the occurrence of the fault, while the other residuals, $\|r_{SMl}\|$, are kept below the corresponding thresholds. In other words, condition (6.18) matches the intuitive idea that a fault can be detected and isolated only if its magnitude overcomes the effect of the uncertainties and disturbances. Of course, (6.18) may result too conservative, especially if the bounds $\bar{\mu}_{s,i}$ and $\mu_{s,i}$ are overestimated; however, it has the merit of showing in a clear and rigorous way how sensitivity to faults may be affected by uncertainties and noise.

6.4.2 Determination of the Healthy Signal

The key point for fault detection and isolation is the design of a suitable Decision Making System (DMS), which, on the basis of the available measurements (physical sensors) and their estimates (virtual sensors), declares the occurrence of a fault, isolates the possible faulty sensor, and outputs an healthy signal. Once a fault is declared and, eventually, isolated, thanks to redundant temperature measurements, the batch can be brought to completion, provided that a suitable voting of the healthy signal is performed. The logic of the *Voter/Monitor* (the subsystem which votes the healthy signal) is described in the following procedure [38] and diagrammatically depicted in Fig. 6.4. As usual, the procedure is based on the assumption that simultaneous faults on different sensors do not occur.

Voter procedure

Step 1. Compute the detection residuals defined in (6.15); then:

(a) If the residuals do not exceed the fixed thresholds (no fault condition), vote the signal given by the average of the two redundant sensors, i.e., the so-called standard duplex measure.

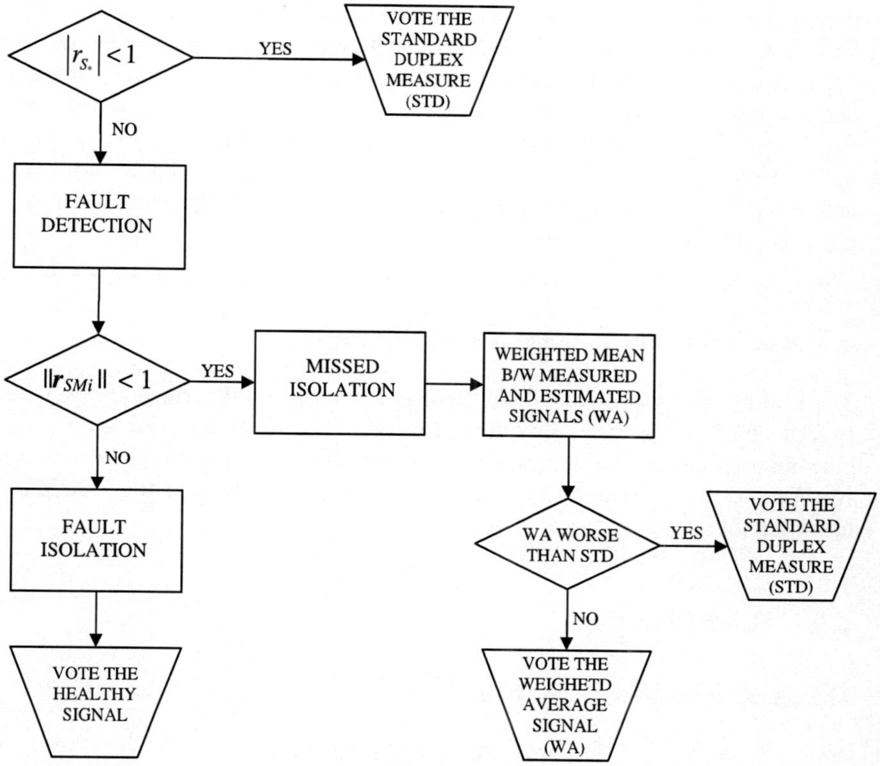

Fig. 6.4 DMS voter logic

(b) If a threshold is exceeded (fault condition), check the isolation residuals $\|r_{SM1}\|$ and $\|r_{SM2}\|$, so as to decide if the faulty signal can be isolated; in this case determine the healthy signal.

Step 2. If, in case (b), fault isolation is not achieved (i.e., both $\|r_{SM1}\|$ and $\|r_{SM2}\|$ are below the respective thresholds), a missed isolation is declared. In this case, the weighted average of the signals provided by the physical and virtual sensors is voted. The weighted average is computed as the arithmetic mean of the measured variable and the output of the sole observer not signaling the occurrence of the fault.

It is of the utmost importance to guarantee that the worst-case performance of the proposed scheme, in terms of voted signal, is not worse than that of a standard duplex measure. Hence, a further elaboration of the residuals is performed. Namely, if the absolute value of the difference between the weighted average (computed at Step 2) and each sensor signal is larger than the difference between the standard duplex measure and the sensor signals, then the standard duplex measure is voted as healthy measure.

Moreover, a rate limiter on the voted signal is adopted, so as to avoid sudden changes of the signals due, for example, to abrupt faults; a step-by-step check verifies if one of the observers is brought to divergence, and, in this case, the related outputs are to be inhibited; if the measurements variance exceeds a threshold value, a fault is signaled: this check is introduced because high-frequency (and zero-mean) additive faults may be filtered by the observer dynamics, causing missed alarms; variations of logical signals are taken into account only if they remain constant for a fixed time window. Noteworthy, a further check on the variation speed of the residuals could be introduced to further enhance detection of abrupt faults, even in the presence of large uncertainties.

6.5 Actuator and Process Fault Diagnosis

The healthy measure, obtained via the diagnostic system described above, is used to feed a bank of observers providing process/actuator fault detection and isolation. One observer detects the occurrence of an actuator or process fault, while the other N_{F} observers, each one corresponding to a fault type, are used for isolation and identification.

6.5.1 Fault Detection

The detection observer has the form

$$\begin{cases} \dot{\widehat{\boldsymbol{x}}}_{\mathrm{a}} = \boldsymbol{A}_{\mathrm{d}}(\boldsymbol{y})\widehat{\boldsymbol{x}}_{\mathrm{a}} + \boldsymbol{b}(\boldsymbol{y}, u) + \boldsymbol{L}_{\mathrm{a}}\widetilde{\boldsymbol{y}}, \\ \widehat{\boldsymbol{y}}_{\mathrm{a}} = \boldsymbol{C}\widehat{\boldsymbol{x}}_{\mathrm{a}}, \end{cases} \tag{6.20}$$

where $\boldsymbol{y}$ is given by the healthy measure voted by the diagnostic system described in Sect. 6.4.2, $\widetilde{\boldsymbol{y}}_{\mathrm{a}} = \boldsymbol{y} - \widehat{\boldsymbol{y}}_{\mathrm{a}}$, and $\boldsymbol{L}_{\mathrm{a}}$ is a matrix gain having the same form of $\boldsymbol{L}_{\mathrm{s}}$. Since the voted healthy measure $\boldsymbol{y}$ can be expressed as in (6.5), the state estimation error dynamics has the form (6.11), i.e.,

$$\begin{cases} \dot{\widetilde{\boldsymbol{x}}}_{\mathrm{a}} = \boldsymbol{A}_{\mathrm{a}}(\boldsymbol{y})\widetilde{\boldsymbol{x}}_{\mathrm{a}} + \boldsymbol{\eta}_{\mathrm{a}}(t, \boldsymbol{x}, u) + \boldsymbol{C}^{\mathrm{T}}\boldsymbol{f}_{\mathrm{a}}(t, \boldsymbol{y}, u), \\ \widetilde{\boldsymbol{y}}_{\mathrm{a}} = \boldsymbol{C}\widetilde{\boldsymbol{x}}_{\mathrm{a}} + \boldsymbol{n}, \end{cases} \tag{6.21}$$

where $\widetilde{\boldsymbol{x}}_{\mathrm{a}} = \boldsymbol{x} - \widehat{\boldsymbol{x}}_{\mathrm{a}}$, $\boldsymbol{A}_{\mathrm{a}} = \boldsymbol{A}_{\mathrm{d}} - \boldsymbol{L}_{\mathrm{a}}\boldsymbol{C}$, and $\boldsymbol{\eta}_{\mathrm{a}} = \boldsymbol{\eta} + \boldsymbol{L}_{\mathrm{a}}\boldsymbol{n}$. Hence, the global asymptotic convergence of the state estimation error, in the absence of faults, uncertainties, and sensor noise (i.e., $\boldsymbol{f}_{\mathrm{a}} = \boldsymbol{0}$, $\boldsymbol{\eta} = \boldsymbol{0}$, and $\boldsymbol{n} = \boldsymbol{0}$), can be established by invoking the same arguments used to prove Theorem 6.1.

Moreover, in the absence of faults and in the presence of uncertainties and disturbances, a bound on the output estimation error can be found in a similar way as in Sect. 6.4, i.e.,

$$\begin{aligned} \|\widetilde{\boldsymbol{y}}_{\mathrm{a}}(t)\| &= \|\boldsymbol{C}\widetilde{\boldsymbol{x}}_{\mathrm{a}}(t) + \boldsymbol{n}(t)\| \leq \|\widetilde{\boldsymbol{x}}_{\mathrm{a}}(t)\| + \|\boldsymbol{n}(t)\| \\ &\leq \left\| \boldsymbol{\Phi}_{\mathrm{a}}(t, t_0)\widetilde{\boldsymbol{x}}_{\mathrm{a}}(t_0) + \int_{t_0}^{t} \boldsymbol{\Phi}_{\mathrm{a}}(t, \zeta)\boldsymbol{\eta}_{\mathrm{a}}\big(\zeta, \boldsymbol{x}(\zeta), u(\zeta)\big)\, d\zeta \right\| + \|\boldsymbol{n}(t)\| \end{aligned}$$

$$\leq \kappa_{\mathrm{a}}\left(\left\|\widetilde{\boldsymbol{x}}_{\mathrm{a}}(t_0)\right\| + \frac{\bar{\eta}_{\mathrm{a}}}{\lambda_{\mathrm{a}}}\right) + \bar{n} = \bar{\mu}_{\mathrm{a}} \quad \forall t \geq t_0, \tag{6.22}$$

where $\bar{\eta}_{\mathrm{a}} = \bar{\eta} + \|\boldsymbol{L}_{\mathrm{a}}\|\bar{n}$, $\boldsymbol{\Phi}_{\mathrm{a}}(t, t_0)$ is state transition matrix corresponding to $\boldsymbol{A}_{\mathrm{a}}$, and κ_{a} and λ_{a} are positive numbers such that $\|\boldsymbol{\Phi}_{\mathrm{a}}(t, t_0)\| \leq \kappa_{\mathrm{a}} e^{-\lambda_{\mathrm{a}}(t-t_0)}$ for all $t \geq t_0$. Hence, the output estimation error of the detection observer keeps bounded in the presence of bounded uncertainties and sensor noise. Moreover, the bound can be, in principle, computed if $\bar{\eta}_{\mathrm{a}}$, $\bar{n}$, κ_{a}, λ_{a}, and $\widetilde{\boldsymbol{x}}_{\mathrm{a}}(t_0)$ are known or, at least, estimated with reasonable accuracy. Of course, the same conclusions drawn in Remark 6.1 on the convergence rate apply to $\widetilde{\boldsymbol{y}}_{\mathrm{a}}(t)$.

A fault is declared when the norm of the residual vector

$$\boldsymbol{r}_{\mathrm{a}} = \frac{\widetilde{\boldsymbol{y}}_{\mathrm{a}}}{\mu_{\mathrm{a}}} \tag{6.23}$$

exceeds a suitably defined threshold. The factor μ_{a}, as usual, is a normalization factor, which can be selected in such a way that threshold can be set to 1. A possible choice for the normalization factor μ_{a} is represented by $\bar{\mu}_{\mathrm{a}}$ in (6.22).

In the presence of a fault occurring at $t = t_{\mathrm{f}}$, the state estimation error can be expressed as

$$\widetilde{\boldsymbol{x}}_{\mathrm{a}}(t) = \boldsymbol{\Phi}_{\mathrm{a}}(t, t_0)\widetilde{\boldsymbol{x}}_{\mathrm{a}}(t_0) + \int_{t_{\mathrm{f}}}^{t} \boldsymbol{\Phi}_{\mathrm{a}}(t, \zeta)\boldsymbol{C}^{\mathrm{T}}\boldsymbol{f}_{\mathrm{a}}\big(\zeta, \boldsymbol{y}(\zeta), \boldsymbol{u}(\zeta)\big)\, d\zeta$$
$$+ \int_{t_0}^{t} \boldsymbol{\Phi}_{\mathrm{a}}(t, \zeta)\boldsymbol{\eta}_{\mathrm{a}}\big(\zeta, \boldsymbol{x}(\zeta), \boldsymbol{u}(\zeta)\big)\, d\zeta \quad \forall t \geq t_{\mathrm{f}}, \tag{6.24}$$

from which the following chain of inequalities can be derived:

$$\begin{aligned}
\|\widetilde{\boldsymbol{y}}_{\mathrm{a}}(t)\| &= \|\boldsymbol{C}\widetilde{\boldsymbol{x}}_{\mathrm{a}}(t) + \boldsymbol{n}(t)\| \\
&\geq \left\| \int_{t_{\mathrm{f}}}^{t} \boldsymbol{C}\boldsymbol{\Phi}_{\mathrm{a}}(t, \zeta)\boldsymbol{C}^{\mathrm{T}}\boldsymbol{f}_{\mathrm{a}}\big(\zeta, \boldsymbol{y}(\zeta), \boldsymbol{u}(\zeta)\big)\, d\zeta \right\| \\
&\quad - \left\| \boldsymbol{C}\boldsymbol{\Phi}_{\mathrm{a}}(t, t_0)\widetilde{\boldsymbol{x}}_{\mathrm{a}}(t_0) + \int_{t_{\mathrm{f}}}^{t} \boldsymbol{C}\boldsymbol{\Phi}_{\mathrm{a}}(t, \zeta)\boldsymbol{\eta}_{\mathrm{a}}\big(\zeta, \boldsymbol{x}(\zeta), \boldsymbol{u}(\zeta)\big)\, d\zeta + \boldsymbol{n}(t) \right\| \\
&\geq \left\| \int_{t_{\mathrm{f}}}^{t} \boldsymbol{C}\boldsymbol{\Phi}_{\mathrm{a}}(t, \zeta)\boldsymbol{C}^{\mathrm{T}}\boldsymbol{f}_{\mathrm{a}}\big(\zeta, \boldsymbol{y}(\zeta), \boldsymbol{u}(\zeta)\big)\, d\zeta \right\| - \bar{\mu}_{\mathrm{a}} \quad \forall t \geq t_{\mathrm{f}}.
\end{aligned}$$

Hence, a sufficient condition for correct detection of the fault is given by

$$\exists t > t_{\mathrm{f}}: \quad \left\| \int_{t_{\mathrm{f}}}^{t} \boldsymbol{C}\boldsymbol{\Phi}_{\mathrm{a}}(t, \zeta)\boldsymbol{C}^{\mathrm{T}}\boldsymbol{f}_{\mathrm{a}}\big(\zeta, \boldsymbol{y}(\zeta), \boldsymbol{u}(\zeta)\big)\, d\zeta \right\| > \bar{\mu}_{\mathrm{a}} + \mu_{\mathrm{a}}, \tag{6.25}$$

since it guarantees that $\|\boldsymbol{r}_{\mathrm{a}}\| = \|\widetilde{\boldsymbol{y}}_{\mathrm{a}}\|/\mu_{\mathrm{a}}$ exceeds 1 at least for a time instant after the occurrence of the fault. Again, condition (6.25) matches the intuitive idea that a process/actuator fault can be detected only if its effect on the estimation error dynamics has a magnitude larger than the effect of the uncertainties. Also, (6.25) may result too conservative if the bounds $\bar{\mu}_{\mathrm{a}}$ and μ_{a} are overestimated.

6.5.2 Fault Isolation and Identification

Once a process/actuator fault has been detected, isolation and identification can be achieved via N_{F} nonlinear adaptive observers. Each observer is designed in such a way to be insensitive to a particular type of fault. In fact, the ith observer (hereafter $i = 1, \dots, N_{\mathrm{F}}$) has the form

$$\begin{cases} \dot{\widehat{\boldsymbol{x}}}_i = \boldsymbol{A}_{\mathrm{d}}(\boldsymbol{y})\widehat{\boldsymbol{x}}_i + \boldsymbol{b}(\boldsymbol{y}, u) + \boldsymbol{L}_{\mathrm{a},i}\widetilde{\boldsymbol{y}}_i + \boldsymbol{C}^{\mathrm{T}}\widehat{\boldsymbol{f}}_{\mathrm{a},i}(t, \boldsymbol{y}, u), \\ \widehat{\boldsymbol{y}}_i = \boldsymbol{C}\widehat{\boldsymbol{x}}_i, \end{cases} \tag{6.26}$$

where $\boldsymbol{L}_{\mathrm{a},i}$ is a gain matrix having the same structure of $\boldsymbol{L}_{\mathrm{a}}$ and $\boldsymbol{L}_{\mathrm{s}}$, $\boldsymbol{y}$ is given by the healthy measure voted by the sensor diagnostic system, $\widetilde{\boldsymbol{y}}_i = \boldsymbol{y} - \widehat{\boldsymbol{y}}_i$, and $\widehat{\boldsymbol{f}}_{\mathrm{a},i}$ is an estimate of the ith fault in $\mathcal{F}_{\mathrm{a}}$ that, in view of (6.8), can be obtained as follows:

$$\widehat{\boldsymbol{f}}_{\mathrm{a},i}(t, \boldsymbol{y}, u) = \boldsymbol{\varphi}_i(t, \boldsymbol{y}, u)\widehat{\boldsymbol{\theta}}_{\mathrm{f},i}, \tag{6.27}$$

where $\widehat{\boldsymbol{\theta}}_{\mathrm{f},i}$ is an estimate of the unknown vector of fault parameters. The adaptive law for $\widehat{\boldsymbol{\theta}}_{\mathrm{f},i}$ is derived by using the Lyapunov synthesis approach (see [5] and Appendix A.6):

$$\dot{\widehat{\boldsymbol{\theta}}}_{\mathrm{f},i} = \gamma_i^{-1}\boldsymbol{\varphi}_i^{\mathrm{T}}(t, \boldsymbol{y}, u)\widetilde{\boldsymbol{y}}_i, \tag{6.28}$$

where γ_i is a positive gain. In order to ensure $\widehat{\boldsymbol{f}}_{\mathrm{a},i} = \boldsymbol{0}$ prior to the detection of the fault, the initial value of $\widehat{\boldsymbol{\theta}}_{\mathrm{f},i}$ could be set to zero, and the parameters update could be activated only after a fault is detected.

In the presence of the ith fault (i.e., $\boldsymbol{f}_{\mathrm{a}} = \boldsymbol{f}_{\mathrm{a},i}$), the state estimation error $\widetilde{\boldsymbol{x}}_i$ of the ith observer (6.26) is given by

$$\begin{cases} \dot{\widetilde{\boldsymbol{x}}}_i = \boldsymbol{A}_{\mathrm{a},i}(\boldsymbol{y})\widetilde{\boldsymbol{x}}_i + \boldsymbol{C}^{\mathrm{T}}\boldsymbol{\varphi}_i(t, \boldsymbol{y}, u)\widetilde{\boldsymbol{\theta}}_{\mathrm{f},i} + \boldsymbol{\eta}_{\mathrm{a},i}(t, \boldsymbol{x}, u), \\ \widetilde{\boldsymbol{y}}_i = \boldsymbol{C}\widetilde{\boldsymbol{x}}_i + \boldsymbol{n}, \end{cases} \tag{6.29}$$

where $\widetilde{\boldsymbol{x}}_i = \boldsymbol{x} - \widehat{\boldsymbol{x}}_i$, $\widetilde{\boldsymbol{\theta}}_{\mathrm{f},i} = \boldsymbol{\theta}_{\mathrm{f},i} - \widehat{\boldsymbol{\theta}}_{\mathrm{f},i}$, $\boldsymbol{A}_{\mathrm{a},i} = \boldsymbol{A}_{\mathrm{d}} - \boldsymbol{L}_{\mathrm{a},i}\boldsymbol{C}$, and $\boldsymbol{\eta}_{\mathrm{a},i} = \boldsymbol{\eta} + \boldsymbol{L}_{\mathrm{a},i}\boldsymbol{n}$. The convergence of the state and parameter estimation errors (in the absence of uncertainties and sensors noise) is stated by the following theorem.

Theorem 6.2 *In the presence of the ith fault* (*i.e.*, $\boldsymbol{f}_{\mathrm{a}} = \boldsymbol{f}_{\mathrm{a},i}$) *and in the absence of uncertainties and sensor noise* (*i.e.*, $\boldsymbol{\eta} = \boldsymbol{0}$ *and* $\boldsymbol{n} = \boldsymbol{0}$), *if the rate constants are bounded as in* (2.32) *and* (2.33), *there exists a set of observer gains such that the state estimation error $\widetilde{\boldsymbol{x}}_i$ of the observer* (6.29) *is globally uniformly convergent to* $\boldsymbol{0}$ *as $t \to \infty$, and the parameter estimation error $\widetilde{\boldsymbol{\theta}}_{\mathrm{f},i}$ is uniformly bounded for every t.*

The proof is based on the arguments in [5] and is reported in Appendix A.6.

Remarkably, in the presence of bounded uncertainties and sensors noise, the boundedness of $\widetilde{\boldsymbol{\theta}}_{\mathrm{f},i}$ is no longer guaranteed. A sufficient condition to achieve the boundedness is given by the persistency of excitation [2, 27] (see Remark 5.1). If the persistency of excitation cannot be met, the update law (6.28) can be modified by adopting the so-called projection operator [22] (see Appendix A.6 for further details). In view of the above remark, the boundedness of $\widetilde{\boldsymbol{\theta}}_{\mathrm{f},i}$ in the presence of bounded uncertainties is assumed hereafter, i.e., $\|\widetilde{\boldsymbol{\theta}}_{\mathrm{f},i}(t)\| \leq \bar{\theta}_{\mathrm{f},i}\ \forall t \geq t_0$.

To achieve fault isolation, the following residuals are computed:

$$\boldsymbol{r}_{\mathrm{a},i} = \frac{\widetilde{\boldsymbol{y}}_i}{\mu_{\mathrm{a},i}}, \quad i = 1, \ldots, N_{\mathrm{F}}, \tag{6.30}$$

where, as usual, $\mu_{\mathrm{a},i}$ are normalization factors selected in such a way to set the thresholds to 1. If the ith fault occurs, the norm of all residuals but $\boldsymbol{r}_{\mathrm{a},i}$ exceeds its threshold.

It is worth remarking that, when a fault occurs which is not included in the N_{F} types considered in the design of the bank of observers, it can be only detected but not isolated and identified.

By using the same arguments developed in previous sections it can be shown that, in the presence of bounded uncertainties and noise, the norm of the output estimation error can be upper bounded as follows:

$$\begin{aligned}
\|\widetilde{\boldsymbol{y}}_i(t)\| &= \|\boldsymbol{C}\widetilde{\boldsymbol{x}}_i(t) + \boldsymbol{n}(t)\| \leq \|\widetilde{\boldsymbol{x}}_i(t)\| + \|\boldsymbol{n}(t)\| \\
&\leq \Bigg\| \boldsymbol{\Phi}_{\mathrm{a},i}(t,t_0)\widetilde{\boldsymbol{x}}_i(t_0) + \int_{t_{\mathrm{f}}}^{t} \boldsymbol{\Phi}_{\mathrm{a},i}(t,\zeta)\boldsymbol{C}^{\mathrm{T}}\boldsymbol{\varphi}_i\big(\zeta, \boldsymbol{y}(\zeta), \boldsymbol{u}(\zeta)\big)\widetilde{\boldsymbol{\theta}}_{\mathrm{f},i}\, d\zeta \\
&\quad + \int_{t_0}^{t} \boldsymbol{\Phi}_{\mathrm{a},i}(t,\zeta)\boldsymbol{\eta}_{\mathrm{a},i}\big(\zeta, \boldsymbol{x}(\zeta), \boldsymbol{u}(\zeta)\big)\, d\zeta \Bigg\| + \|\boldsymbol{n}(t)\| \\
&\leq \kappa_{\mathrm{a},i} e^{-\lambda_{\mathrm{a},i}(t-t_0)} \|\widetilde{\boldsymbol{x}}_i(t_0)\| + \int_{t_{\mathrm{f}}}^{t} \kappa_{\mathrm{a},i}\bar{\varphi}_i\bar{\theta}_{\mathrm{f},i} e^{-\lambda_{\mathrm{a},i}(t-\zeta)}\, d\zeta \\
&\quad + \int_{t_0}^{t} \kappa_{\mathrm{a},i}\bar{\eta}_{\mathrm{a},i} e^{-\lambda_{\mathrm{a},i}(t-\zeta)}\, d\zeta + \bar{n} \\
&\leq \kappa_{\mathrm{a},i} e^{-\lambda_{\mathrm{a},i}(t-t_0)} \|\widetilde{\boldsymbol{x}}_i(t_0)\| + \frac{\kappa_{\mathrm{a},i}\bar{\varphi}_i\bar{\theta}_{\mathrm{f},i}}{\lambda_{\mathrm{a},i}}\big(1 - e^{-\lambda_{\mathrm{a},i}(t-t_{\mathrm{f}})}\big) \\
&\quad + \frac{\kappa_{\mathrm{a},i}\bar{\eta}_{\mathrm{a},i}}{\lambda_{\mathrm{a},i}}\big(1 - e^{-\lambda_{\mathrm{a},i}(t-t_0)}\big) + \bar{n} \\
&\leq \kappa_{\mathrm{a},i}\left(\|\widetilde{\boldsymbol{x}}_i(t_0)\| + \frac{\bar{\varphi}_i\bar{\theta}_{\mathrm{f},i}}{\lambda_{\mathrm{a},i}} + \frac{\bar{\eta}_{\mathrm{a},i}}{\lambda_{\mathrm{a},i}} \right) + \bar{n} = \bar{\mu}_{\mathrm{a},i},
\end{aligned} \tag{6.31}$$

where $\bar{\eta}_{\mathrm{a},i} = \bar{\eta} + \|\boldsymbol{L}_{\mathrm{a},i}\|\bar{n}$, $\boldsymbol{\Phi}_{\mathrm{a},i}(t,t_0)$ is the state transition matrix corresponding to $\boldsymbol{A}_{\mathrm{a},i}$, and $\kappa_{\mathrm{a},i}$ and $\lambda_{\mathrm{a},i}$ are suitable positive constants. If $\bar{\mu}_{\mathrm{a},i}$ can be estimated, it represents a possible choice for the normalization factors $\mu_{\mathrm{a},i}$. The above inequality ensures that the output estimation errors of the ith observer keep bounded, in the presence of bounded uncertainties and sensors noise, provided that $\boldsymbol{\varphi}_i\widetilde{\boldsymbol{\theta}}_{\mathrm{f},i}$ is bounded.

In the presence of the same fault, the dynamics of the state estimation error $\widetilde{\boldsymbol{x}}_l$ of the lth observer ($l \neq i$) is given by

$$\begin{cases} \dot{\widetilde{\boldsymbol{x}}}_l = \boldsymbol{A}_{\mathrm{a},l}(\boldsymbol{y})\widetilde{\boldsymbol{x}}_l + \boldsymbol{C}^{\mathrm{T}}\big(\boldsymbol{\varphi}_i(t,\boldsymbol{y},\boldsymbol{u})\boldsymbol{\theta}_{\mathrm{f},i} - \boldsymbol{\varphi}_l(t,\boldsymbol{y},\boldsymbol{u})\widehat{\boldsymbol{\theta}}_{\mathrm{f},l}\big) + \boldsymbol{\eta}_{\mathrm{a},l}(t,\boldsymbol{x},\boldsymbol{u}), \\ \widetilde{\boldsymbol{y}}_l = \boldsymbol{C}\widetilde{\boldsymbol{x}}_l + \boldsymbol{n}. \end{cases} \tag{6.32}$$

The state estimation error $\widetilde{\boldsymbol{x}}_l$ is then given by

$$\begin{aligned} \widetilde{\boldsymbol{x}}_l(t) &= \boldsymbol{\Phi}_{\mathrm{a},l}(t,t_0)\widetilde{\boldsymbol{x}}_l(t_0) \\ &\quad + \int_{t_\mathrm{f}}^{t} \boldsymbol{\Phi}_{\mathrm{a},l}(t,\zeta)\boldsymbol{C}^{\mathrm{T}}\big(\boldsymbol{\varphi}_i\big(\zeta,\boldsymbol{y}(\zeta),\boldsymbol{u}(\zeta)\big)\boldsymbol{\theta}_{\mathrm{f},i} - \boldsymbol{\varphi}_l\big(\zeta,\boldsymbol{y}(\zeta),\boldsymbol{u}(\zeta)\big)\widehat{\boldsymbol{\theta}}_{\mathrm{f},l}\big)\,d\zeta \\ &\quad + \int_{t_0}^{t} \boldsymbol{\Phi}_{\mathrm{a},l}(t,\zeta)\boldsymbol{\eta}_{\mathrm{a},l}\big(\zeta,\boldsymbol{x}(\zeta),\boldsymbol{u}(\zeta)\big)\,d\zeta \quad \forall t \geq t_\mathrm{f}, \end{aligned} \tag{6.33}$$

from which the following inequality can be derived for all $t \geq t_\mathrm{f}$:

$$\begin{aligned} \|\widetilde{\boldsymbol{y}}_l(t)\| \geq \bigg\| \int_{t_\mathrm{f}}^{t} & \boldsymbol{C}\boldsymbol{\Phi}_{\mathrm{a},l}(t,\zeta)\boldsymbol{C}^{\mathrm{T}}\big(\boldsymbol{\varphi}_i\big(\zeta,\boldsymbol{y}(\zeta),\boldsymbol{u}(\zeta)\big)\boldsymbol{\theta}_{\mathrm{f},i} \\ & - \boldsymbol{\varphi}_l\big(\zeta,\boldsymbol{y}(\zeta),\boldsymbol{u}(\zeta)\big)\widehat{\boldsymbol{\theta}}_{\mathrm{f},l}\big)\,d\zeta \bigg\| - \bar{\mu}_{\mathrm{a},l}. \end{aligned}$$

Hence, a sufficient condition for isolability for the ith type of process/actuators faults is given by the two inequalities,

$$\begin{aligned} &\exists t > t_\mathrm{f}: \\ &\bigg\| \int_{t_\mathrm{f}}^{t} \boldsymbol{C}\boldsymbol{\Phi}_{\mathrm{a},l}(t,\zeta)\boldsymbol{C}^{\mathrm{T}}\big(\boldsymbol{\varphi}_i\big(\zeta,\boldsymbol{y}(\zeta),\boldsymbol{u}(\zeta)\big)\boldsymbol{\theta}_{\mathrm{f},i} - \boldsymbol{\varphi}_l\big(\zeta,\boldsymbol{y}(\zeta),\boldsymbol{u}(\zeta)\big)\widehat{\boldsymbol{\theta}}_{\mathrm{f},l}\big)\,d\zeta \bigg\| \\ &> \bar{\mu}_{\mathrm{a},l} + \mu_{\mathrm{a},l} \\ &\forall l = 1,\dots,N_\mathrm{F} \ (l \neq i) \end{aligned} \tag{6.34}$$

and

$$\|\widetilde{\boldsymbol{y}}_i(t)\| \leq \bar{\mu}_{\mathrm{a},i} \quad \forall t > t_\mathrm{f}, \tag{6.35}$$

which guarantees that the all the residuals $\|\boldsymbol{r}_{\mathrm{a},l}\| = \|\widetilde{\boldsymbol{y}}_l\|/\mu_{\mathrm{a},l}$ ($l \neq i$) exceed 1 at least for a time instant, while the ith residual keeps below its threshold.

The left-hand side of inequality (6.34) can be interpreted as a measure of the capability of the lth isolation observer to estimate the ith fault. It can be argued that such a capability may depend on different factors, such as structural similarities of two distinct faults in $\mathcal{F}_\mathrm{a}$ and the dynamics each isolation observer. In Sect. 6.7.2, a case study illustrating this behavior in practical case is developed.

6.6 Decoupling Sensor Faults from Process and Actuator Faults

In order to make the observer (6.10) insensitive to process/actuator faults, the following modified dynamics can be adopted:

$$\begin{cases} \dot{\widehat{x}}_{\mathrm{SM}i} = A_{\mathrm{d}}(y_{\mathrm{SM}i})\widehat{x}_{\mathrm{SM}i} + b(y_{\mathrm{SM}i}, u) + L_{\mathrm{s}}\widetilde{y}_{\mathrm{SM}i} + C^{\mathrm{T}}\widehat{f}_{\mathrm{a}}(t, y, u), \\ \widehat{y}_{\mathrm{SM}i} = C\widehat{x}_{\mathrm{SM}i}, \end{cases} \tag{6.36}$$

where y is the healthy measure voted according to the procedure in Sect. 6.4.2, and $\widehat{f}_{\mathrm{a}}(t, y, u)$ is an estimate of the isolated process/actuator fault. In other words, if the ith process/actuator fault has been detected and isolated, then $\widehat{f}_{\mathrm{a}}(t, y, u) = \widehat{f}_{\mathrm{a},i}(t, y, u)$. Indeed, by invoking the same arguments used to establish (6.31), it can be shown that adoption of the form (6.36) instead of (6.10) guarantees that the output estimation error, $\widetilde{y}_{\mathrm{SM}i}$, is only marginally influenced by the process/actuator fault, provided that a bounded (possibly small) error on the fault estimation is achieved.

6.7 Case Study: Fault Diagnosis

The effectiveness of the proposed approach has been tested in simulation by considering a jacketed batch reactor in which the phenol–formaldehyde reaction presented in Chap. 2 takes place. The complete system of differential equations given by the 13 mass balances presented in Sect. 2.4 has been simulated in the MATLAB/SIMULINK© environment.

The same assumptions in Chap. 5 on the experimental setup have been done. The reactor parameters and the initial conditions for the reactant concentrations and the temperatures of the vessel and the jacket are reported in Table 5.1. The model-based temperature controller proposed in Chap. 5 is adopted. Finally, both in the reactor vessel and in the jacket, duplicated temperature sensors have been considered.

In order to design the bank of observers and the controller, the reduced model (3.57) identified in Chap. 3 with first-order kinetics has been considered. Moreover, a 5% estimation error on the parameter θ has been assumed, i.e., $\theta^\star = 0.95 \cdot \theta$.

Matrix $A_{\mathrm{d}}(y)$ has the following form:

$$A_{\mathrm{d}}(y) = \begin{bmatrix} -k_{\mathrm{c}1}(T_{\mathrm{r}}) & 0 & 0 & 0 & 0 \\ k_{\mathrm{c}1}(T_{\mathrm{r}}) & -k_{\mathrm{c}2}(T_{\mathrm{r}}) & 0 & 0 & 0 \\ 0 & k_{\mathrm{c}2}(T_{\mathrm{r}}) & -k_{\mathrm{c}3}(T_{\mathrm{r}}) & 0 & 0 \\ a_1(T_{\mathrm{r}}) & a_2(T_{\mathrm{r}}) & a_3(T_{\mathrm{r}}) & -\alpha_{\mathrm{r}}\theta^\star & \alpha_{\mathrm{r}}\theta^\star \\ 0 & 0 & 0 & \alpha_{\mathrm{j}}\theta^\star & -\alpha_{\mathrm{j}}\theta^\star \end{bmatrix},$$

where the quantities a_i $(i = 1, 2, 3)$ are those defined in (5.52).

It is worth remarking that the above reduced model has been adopted so as to emulate the presence of modeling errors with a twofold nature, namely:

- structural uncertainties, since the model used to build the observers is based on a simplified reaction network, where both the reactions and the involved chemical species have been lumped into a reduced set of reactions and species

- parametric uncertainties, since the identified model parameters are affected by unavoidable errors.

The gain matrices of all the diagnostic observers have been set as follows:

$$L_s = L_a = L_{a,i} = \begin{bmatrix} 1 \cdot 10 & 0 \\ 5 \cdot 10^2 & 0 \\ 1 \cdot 10^{-1} & 0 \\ 1 & 0 \\ 0 & 1 \end{bmatrix}. \tag{6.37}$$

The normalization factors $\mu_{s,*}$ $(* = s, r)$, $\mu_{s,i}$ $(i = 1, 2)$, μ_a, and $\mu_{a,i}$ $(i = 1, \dots, N_F)$ have been chosen equal to 0.15; this value has been determined by measuring the maximum values of the estimation errors of the corresponding observers in healthy condition.

6.7.1 Simulation Results: Sensor Faults

In the simulations, the following classes of faults on the temperature sensors have been considered:

- *abrupt switches to zero* of the measured signal
- *slow drifts*, i.e., a linearly increasing signal is added to the measured data
- *abrupt constant biases*, i.e., a step disturbance is added to the measured data
- *abrupt freezing* of the measured signal, i.e., the measured signal is frozen at its value taken at a certain time instant; and
- *increasing noise*, i.e., a Gaussian noise with increasing variance is added to the measured data.

Various simulations with the above kind of faults occurring at different time instants have been done, and the obtained results show that the proposed diagnostic scheme has been able to detect and isolate all the simulated sensor faults. In detail, Figs. 6.5 to 6.12 show the voted measures and the norms of both detection and isolation residuals in the presence of different sensor faults.

Figures 6.5 and 6.6 are referred to a *slow drift* of the output of sensor $S_{j,2}$, i.e., a linearly increasing signal, with a 10^{-3} K s^{-1} rate of change, is added to the measured variable for $t \geq t_f = 9000$ s. It can be recognized that the fault is detected a few time instants after the occurrence, while it is isolated about 2000 s after t_f. This is due to the slow time evolution of the fault; it can be argued that, in the first 2000 s after the occurrence of the fault, its effect is quite negligible and/or almost totally compensated by the observers. In order to reduce the isolation time, the normalization factors could be reduced, at the expense of an increased probability of false alarms. Moreover, Fig. 6.5 shows that the voted measure is the mean value of the measured and estimated data until the isolation is performed, and then it switches to the value of the healthy sensor ($S_{j,1}$).

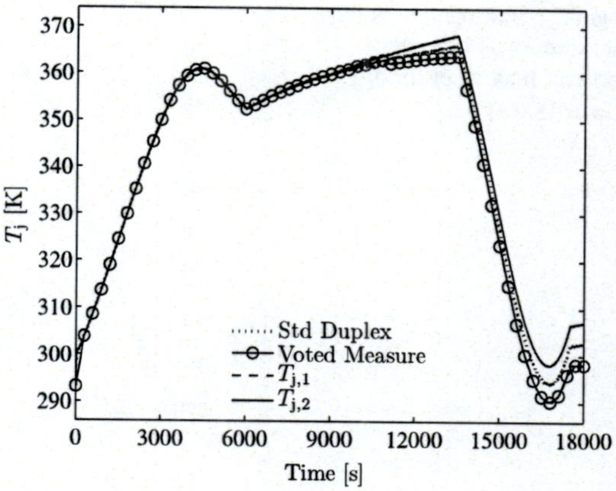

Fig. 6.5 Sensor fault: voted measure for T_j (slow drift at sensor $S_{j,2}$, $t_f = 9000$ s)

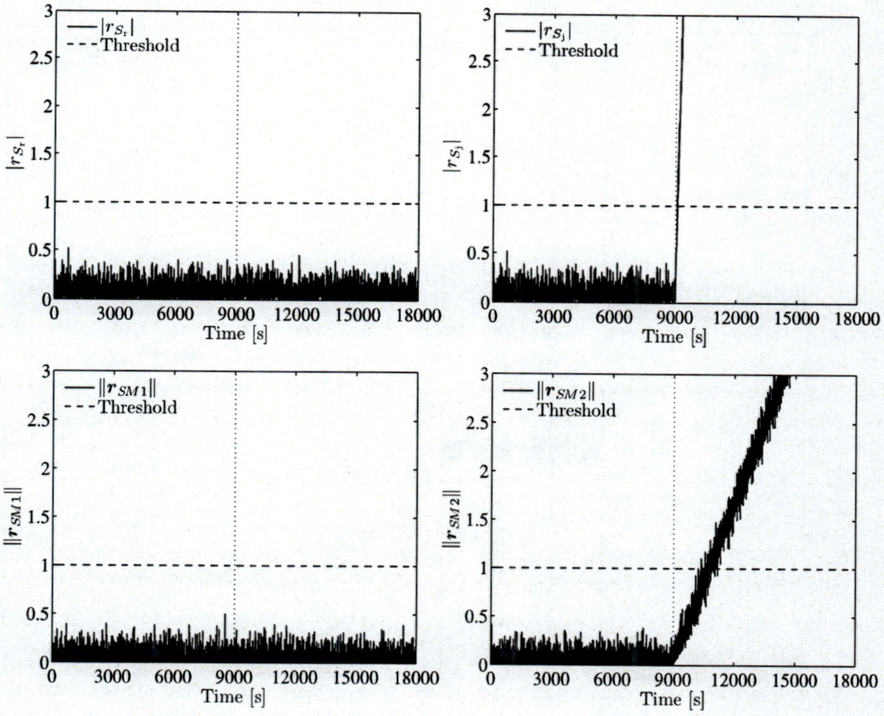

Fig. 6.6 Sensor fault: detection and isolation residuals (slow drift at sensor $S_{j,2}$, $t_f = 9000$ s)

Figures 6.7 and 6.8 report the obtained results in the presence of an *abrupt constant bias*, with an amplitude of 10 K added to the output of sensor $S_{r,1}$, starting at time $t_f = 10\,000$ s.

Figures 6.9 and 6.10 show the results obtained when the measured signal of sensor $S_{j,1}$ has been frozen at its value taken at $t_f = 3000$ s. Since these last two cases involve abrupt faults, detection and isolation are practically contemporary.

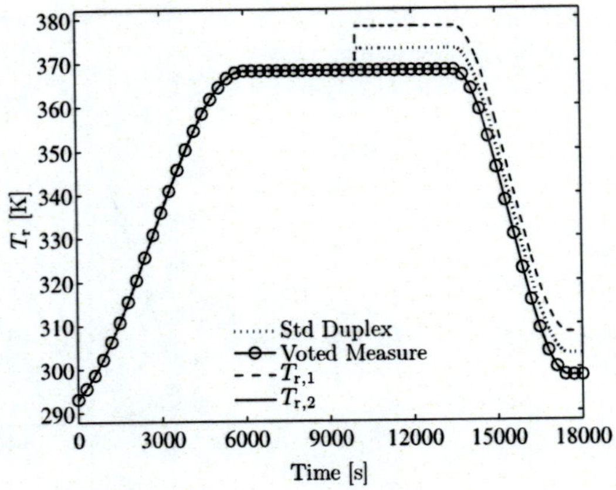

Fig. 6.7 Sensor fault: voted measure for T_r (abrupt constant bias at sensor $S_{r,1}$, $t_f = 10\,000$ s)

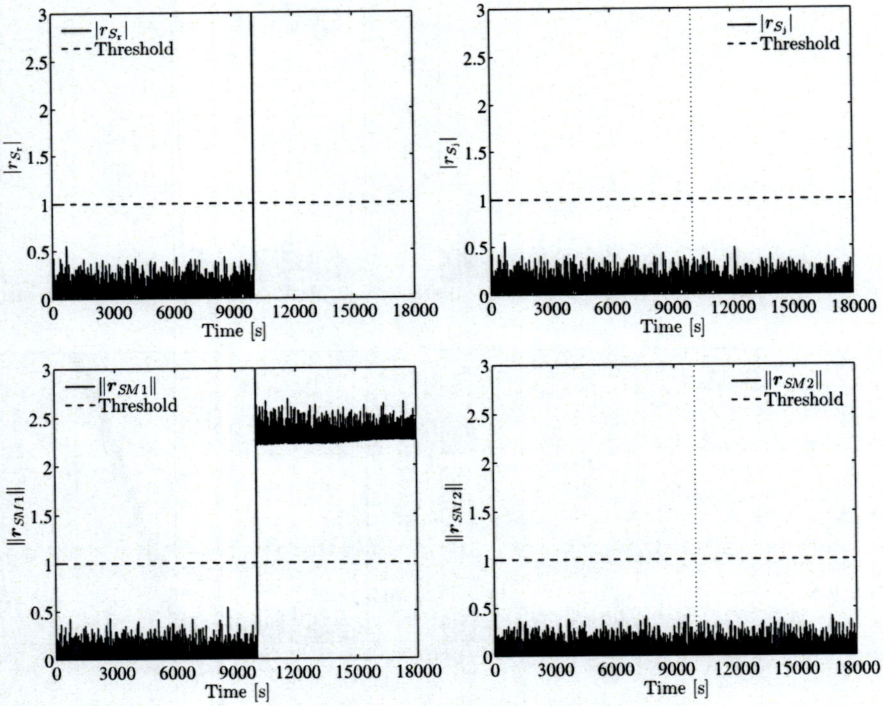

Fig. 6.8 Sensor fault: detection and isolation residuals (abrupt constant bias at sensor $S_{r,1}$, $t_f = 10\,000$ s)

Finally, Figs. 6.11 and 6.12 show the results obtained for an *increasing noise* added to the output of the sensor $S_{r,1}$; namely, a white noise with zero mean and increasing variance has been added to the sensor output starting at time 12 000 s. In this case, the delay for the isolation is about 100 s.

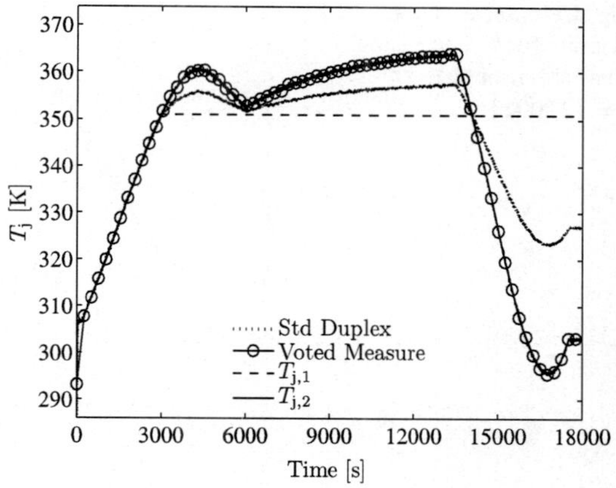

Fig. 6.9 Sensor fault: voted measure for T_j (abrupt freezing at sensor $S_{j,1}$, $t_f = 3000$ s)

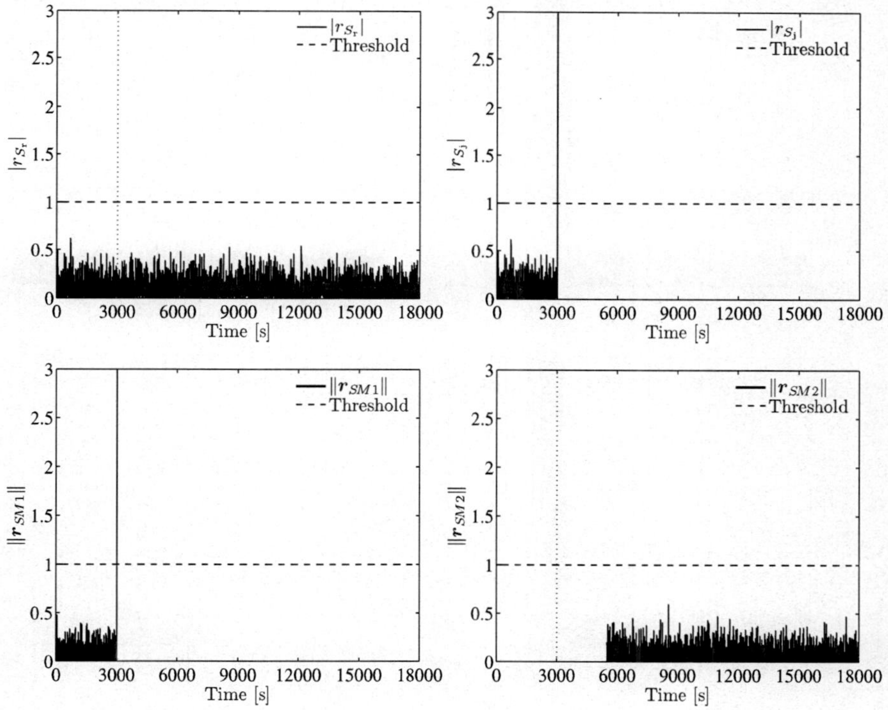

Fig. 6.10 Sensor fault: detection and isolation residuals (abrupt freezing at sensor $S_{j,1}$, $t_f = 3000$ s)

It can be easily recognized that all the faults have been correctly detected and identified. A wide simulation campaign showed that only when an abrupt freezing on reactor temperature sensors (i.e., $S_{r,1}$ or $S_{r,2}$) occurs during the isothermal phase, the fault can be detected but not isolated. Finally, the measure voted by the DMS results to be always more accurate than the standard duplex measure.

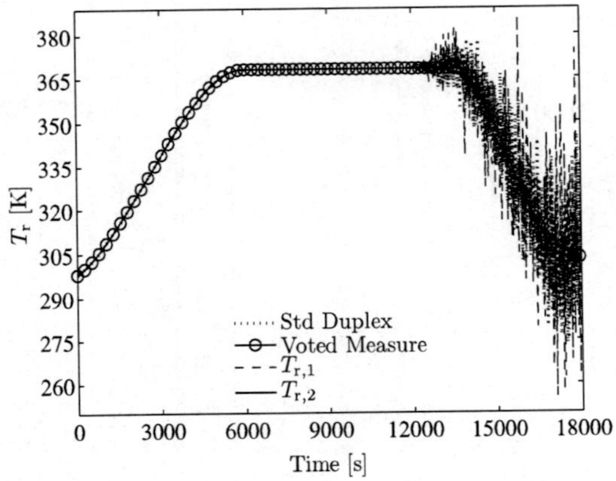

Fig. 6.11 Sensor fault: voted measure for T_j (increasing noise at sensor $S_{r,1}$, $t_f = 12\,000$ s)

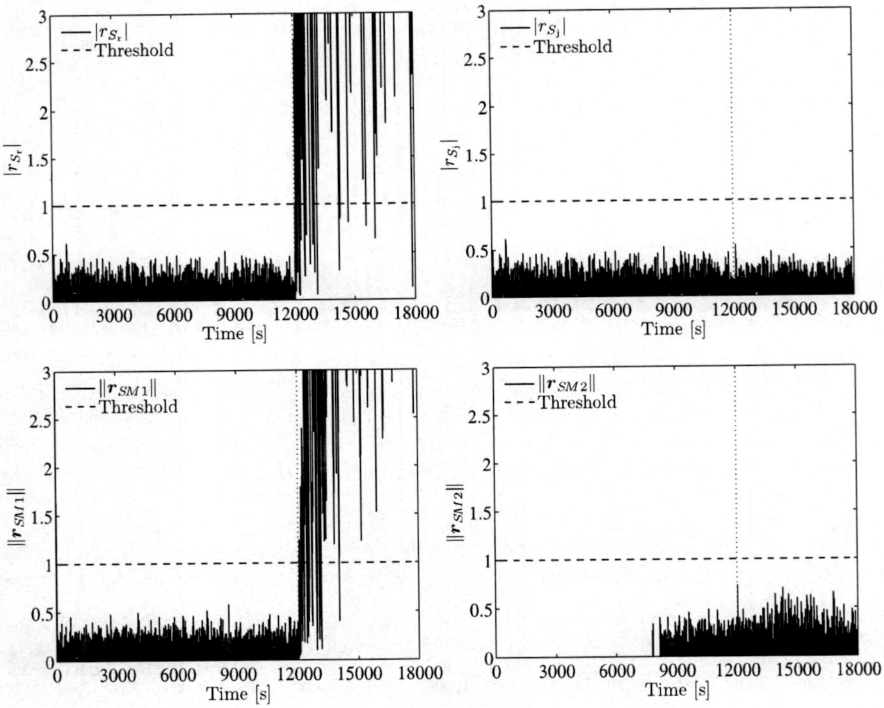

Fig. 6.12 Sensor fault: detection and isolation residuals (increasing noise at sensor $S_{r,1}$, $t_f = 12\,000$ s)

6.7.2 Simulation Results: Process and Actuator Faults

Three classes of actuator and process faults have been considered in the simulations.

A variation of the heat transfer coefficient between the reactor and the jacket (fault type 1) has been modeled as a time-varying bias $\Delta U(t)$, superimposed to the nominal value of the heat transfer coefficient $U^\star$, i.e.,

$$\begin{cases} \Delta U(t) = 0 & \text{if } t < t_\text{f}, \\ \Delta U(t) = \delta_U(1 - e^{-(t-t_\text{f})/\tau_U}) & \text{if } t \geq t_\text{f}, \end{cases} \tag{6.38}$$

where t_f is the fault time, τ_U is a time constant setting the fault rate, and δ_U is the magnitude of the fault. Hence, the effect of the fault can be modeled as

$$\boldsymbol{f}_{\text{a},1}(\boldsymbol{y}) = \boldsymbol{\varphi}_1(\boldsymbol{y})\theta_{\text{f},1}, \tag{6.39}$$

where

$$\theta_{\text{f},1} = \delta_U, \qquad \boldsymbol{\varphi}_1(\boldsymbol{y}) = \begin{bmatrix} -\alpha_\text{r} S(y_1 - y_2) \\ \alpha_\text{j} S(y_1 - y_2) \end{bmatrix}.$$

A fault affecting the thermal insulation of the cooling jacket (fault type 2) results in an additive term Δ_j on the energy balance of the jacket, i.e.,

$$\begin{cases} \Delta_\text{j}(t) = 0 & \text{if } t < t_\text{f}, \\ \Delta_\text{j}(t) = \delta_\text{j}(1 - e^{-(t-t_\text{f})/\tau_\text{j}})(y_2 - T_\text{a}) & \text{if } t \geq t_\text{f}, \end{cases} \tag{6.40}$$

where τ_j is a time constant setting the fault rate, δ_j is the magnitude of the fault, and T_a is the external environment temperature. The effect of the fault can be modeled as follows:

$$\boldsymbol{f}_{\text{a},2}(\boldsymbol{y}) = \boldsymbol{\varphi}_2(\boldsymbol{y})\theta_{\text{f},2} \tag{6.41}$$

with

$$\theta_{\text{f},2} = \delta_\text{j}, \qquad \boldsymbol{\varphi}_2(\boldsymbol{y}) = \begin{bmatrix} 0 \\ (y_2 - T_\text{a}) \end{bmatrix}.$$

A fault affecting the cooling system (fault type 3) is modeled as an additive term $\Delta u(t)$ on the commanded cooling fluid temperature computed by the controller, i.e.,

$$\begin{cases} \Delta u(t) = 0 & \text{if } t < t_\text{f}, \\ \Delta u(t) = \delta_u(1 - e^{-(t-t_\text{f})/\tau_u}) & \text{if } t \geq t_\text{f}, \end{cases} \tag{6.42}$$

where τ_u is a time constant setting the fault rate, and δ_u is the magnitude of the fault. Hence,

$$\boldsymbol{f}_{\text{a},3} = \boldsymbol{\varphi}_3\theta_{\text{f},3} \tag{6.43}$$

with

$$\theta_{\text{f},3} = \delta_u, \qquad \boldsymbol{\varphi}_3 = \begin{bmatrix} 0 \\ \beta_\text{j} \end{bmatrix}.$$

Hence, a bank of three isolation observers has been designed; the gain matrices $\boldsymbol{L}_{\text{a},i}$ have been set as in (6.37), while the gains γ_i setting the parameter update rate have been set, via a trial-and-error procedure, to $\gamma_1 = 0.1$, $\gamma_2 = 1000$, and $\gamma_3 = 0.03$.

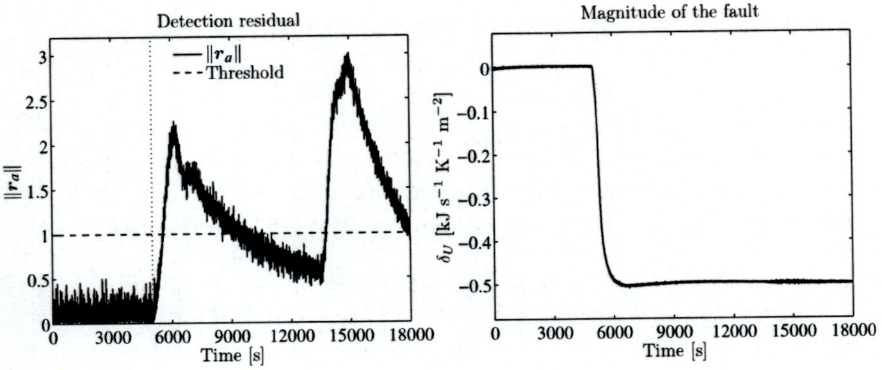

Fig. 6.13 Process fault (type 1): detection residual and estimate of the magnitude of the fault

Fig. 6.14 Process fault (type 1): isolation residuals

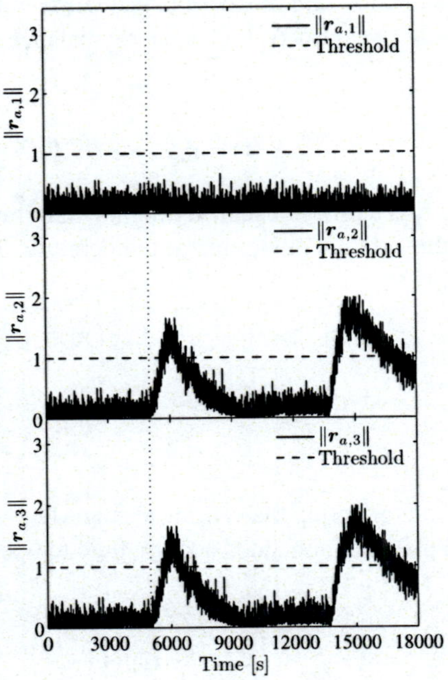

Figures 6.13 and 6.14 report the results obtained when a fault of type 1 has been simulated, with $\tau_U = 300$ s, $\delta_U = -0.5$ kJ s^{-1} K^{-1} m^{-2}, and $t_\mathrm{f} = 5000$ s. Figure 6.13 shows that the fault is detected and identified, i.e., its magnitude is correctly estimated; moreover, Fig. 6.14 shows that only the residuals output of the first observer remains always below the threshold, whereas the other two residuals exceed their corresponding thresholds a few minutes after t_f, thus achieving fault isolation.

Figures 6.15 and 6.16 report the results obtained in the presence of a fault of type 2, with $\tau_\mathrm{j} = 600$ s, $\delta_\mathrm{j} = 0.02$ s^{-1}, $T_\mathrm{a} = 293$ K, and $t_\mathrm{f} = 10\,000$ s. As shown in Fig. 6.15, the fault is clearly detected, and the second observer achieves a fairly accurate fault identification. Figure 6.16 shows the residuals computed by the isolation

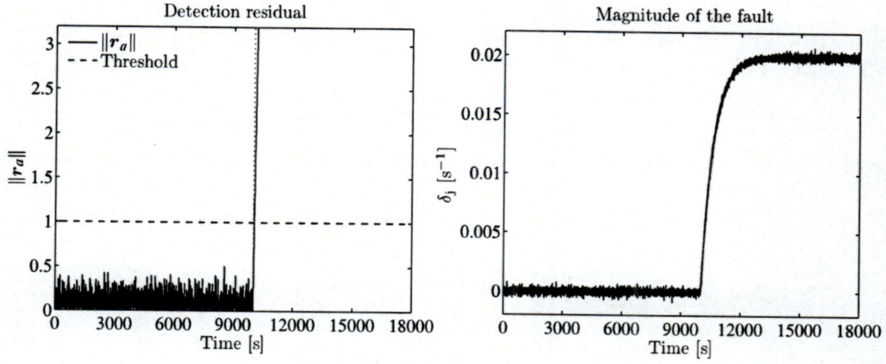

Fig. 6.15 Process fault (type 2): detection residual and estimate of the magnitude of the fault

Fig. 6.16 Process fault (type 2): isolation residuals

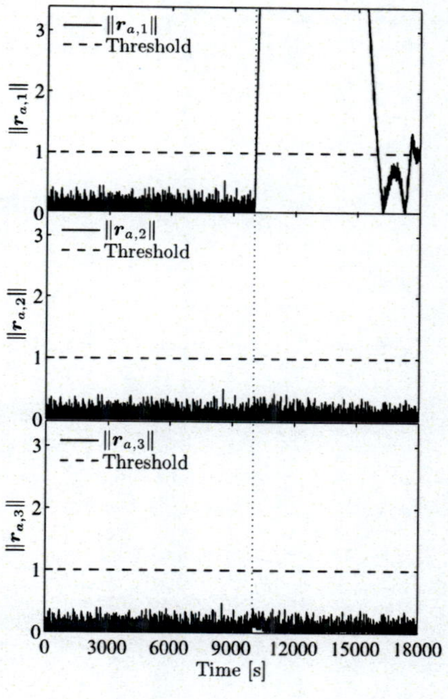

observers; it can be recognized that only the norm of the first residual exceeds the threshold, whereas the other always remain below their corresponding thresholds; hence, the fault cannot be isolated in this case.

Figures 6.17 and 6.18 report the results obtained in the presence of a fault of type 3, with $\tau_u = 600$ s, $\delta_u = 10$ K, and $t_f = 14000$ s. Figure 6.17 shows that the fault has been detected and its magnitude correctly estimated, while isolation residuals in Fig. 6.18 show that fault isolation is achieved as well.

The missed isolation of the second fault can be explained in view of inequality (6.34). Indeed, it can be noticed that fault types 2 and 3 are similar, in the sense the $\boldsymbol{\varphi}_2$ and $\boldsymbol{\varphi}_3$ have the same vector structure; moreover, the gain γ_2 is larger than γ_3.

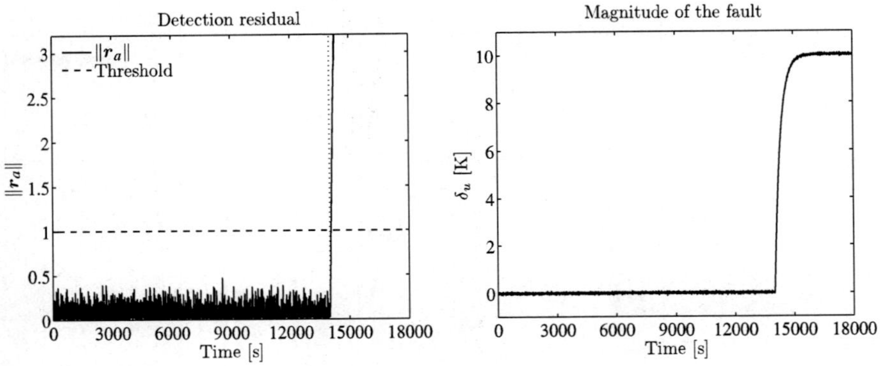

Fig. 6.17 Actuator fault (type 3): detection residual and estimate of the magnitude of the fault

Fig. 6.18 Actuator fault (type 3): isolation residuals

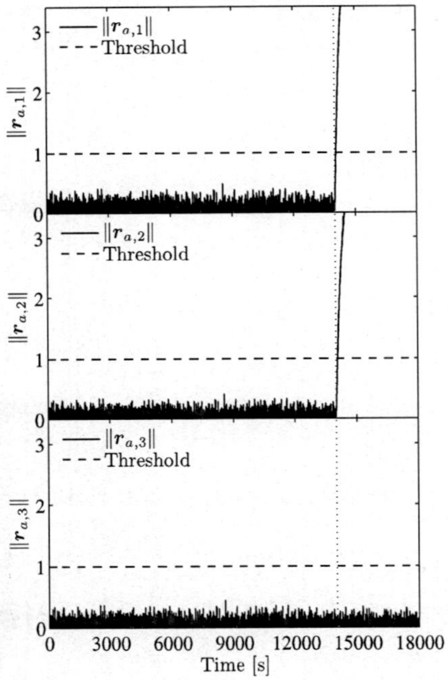

Hence, it can be argued that in the third isolation observer an accurate estimate of the second fault type is achieved. On the contrary, the second observer is not able to estimate the third fault.

6.7.3 Simulation Results: Sensor and Actuator Faults

Finally, a simulation has been carried out in which both a sensor and an actuator fault occur during the same batch run. First, the output of sensor $S_{r,2}$ suddenly *switches*

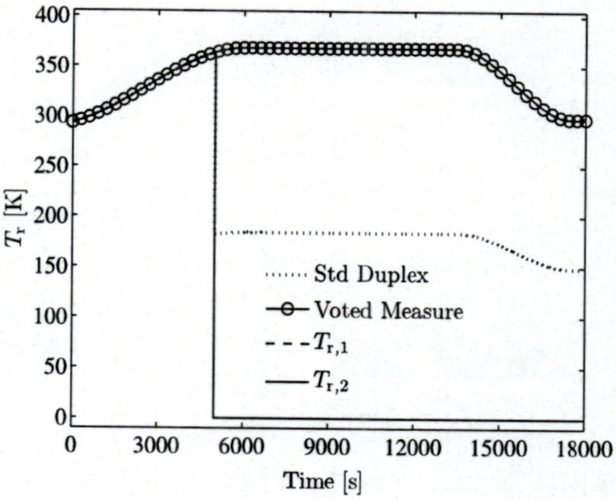

Fig. 6.19 Sensor $S_{r,2}$ and cooling system faults: voted measure of T_r

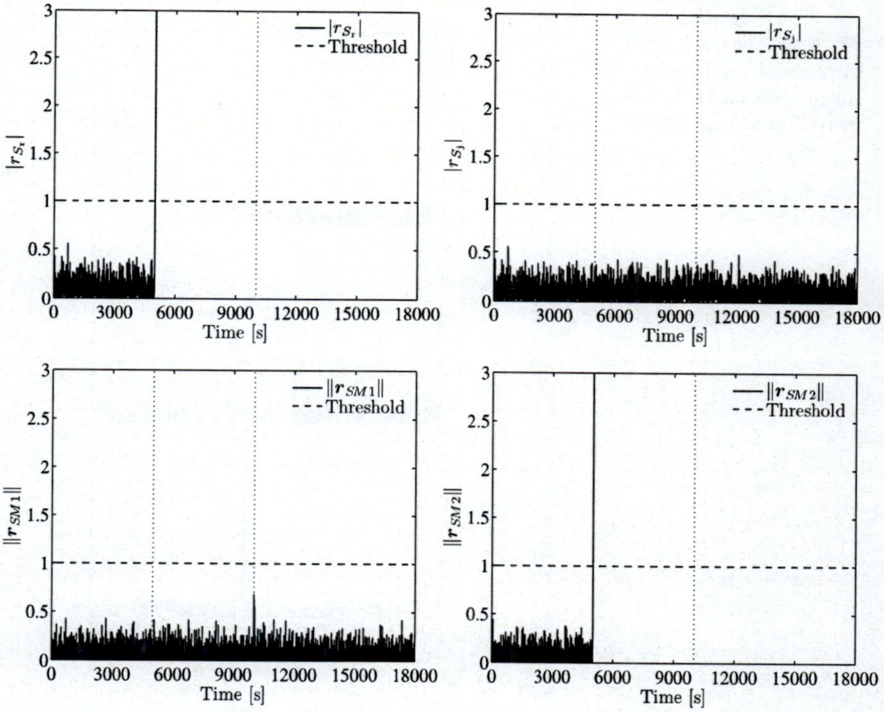

Fig. 6.20 Sensor $S_{r,2}$ and cooling system faults: detection and isolation residuals for the sensor fault

to zero at $t_{f,s} = 5000$ s; then, an actuator fault (fault type 2) with $\delta_j = 0.03\ s^{-1}$ and $\tau_j = 60$ s, occurs at $t_{f,a} = 10\,000$ s. The DMS is able to vote the healthy measure of temperature T_r (Fig. 6.19). Figure 6.20 shows that the residuals $\boldsymbol{r}_{SMi}$ $(i = 1, 2)$, r_{S_r}, and r_{S_j} are able to detect and isolate the sensor fault. Figures 6.21 and 6.22 show the detection and isolation residuals and the estimate of the fault. It can be noticed that

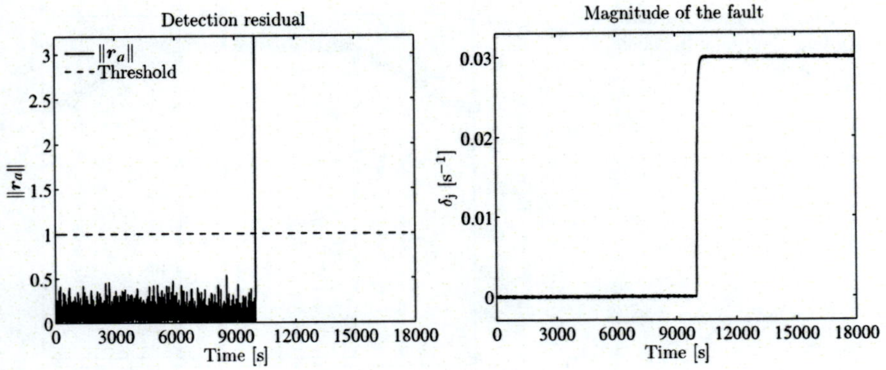

Fig. 6.21 Sensor $S_{r,2}$ and thermal insulation faults: detection residual and estimate of the magnitude of the thermal insulation fault

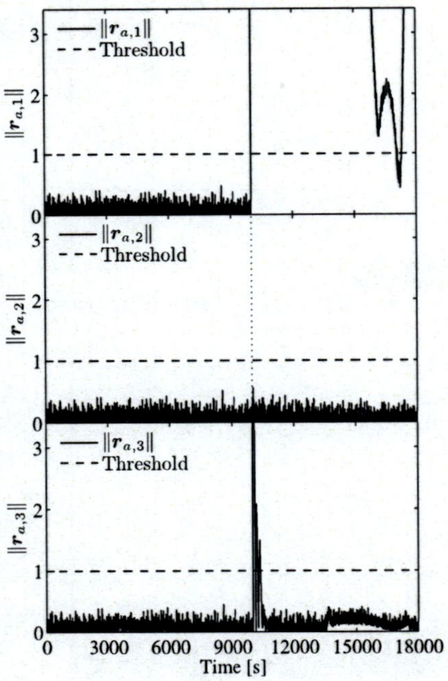

Fig. 6.22 Sensor $S_{r,2}$ and thermal insulation faults: isolation residuals for the thermal insulation fault

residuals r_{SMi}, r_{S_r}, and r_{S_j} are insensitive to the actuator fault, while residuals r_l ($l = 1, 2, 3$) are insensitive to the sensor fault, i.e., the effects of sensor and actuator faults have been correctly decoupled by adopting the observers (6.36) presented in Sect. 6.6.

It is worth noticing that, in this case, the fault affecting the thermal insulation of the jacket (fault type 2) is correctly isolated. This can be explained by noticing that the fault rate is much larger than in the previous case (i.e., τ_j is ten times smaller); hence, the estimate computed by the third observer is not able to track accurately the fault profile during its rising edge, and thus the third residual exceeds the threshold until the observer estimation error is recovered.

6.8 Conclusions

In this chapter an integrated approach to fault diagnosis for chemical batch reactors has been developed. The approach is based on the so-called analytical redundancy, i.e., a model of the system is adopted to design the diagnostic observers needed to achieve fault detection, isolation, and identification. A limited degree of hardware redundancy is assumed as well, since duplicated temperature sensors are considered. The use of duplicated sensor may be justified by the high level of risk of strongly exothermic reactive systems; moreover, thanks to the duplex sensor architecture, the batch can be brought to its completion in the case of sensor faults.

The approach is able to detect, isolate, and identify a wide class of failures of sensors, actuators, and process. Sufficient conditions for residuals convergence, detectability, and isolability of faults have been derived. In detail, detection is guaranteed under mild assumptions on the magnitude of model uncertainties and disturbances, whereas correct isolation may not be achieved if multiple faults of the same nature (i.e., sensor faults and process/actuator faults) occur during the same batch operation.

In the case study, the adaptive model-based approach is designed on the basis of a reduced model of the phenol–formaldehyde reaction introduced in Chap. 2. Noticeably, the results show that the fault diagnosis scheme achieves very good performance even when a strongly simplified mathematical model of the reactive system is adopted for the design.

References

1. E. Alcorta Garcia and P.M. Frank. Deterministic nonlinear observer-based approaches to fault diagnosis: a survey. *Control Engineering Practice*, 5(5):663–670, 1997.
2. K.J. Aström and B. Wittenmark. *Adaptive Control, 2nd Edition*. Addison-Wesley, Reading, 1995.
3. M. Basseville and I.V. Nikiforov. *Detection of Abrupt Changes—Theory and Application*. Information and System Sciences Series. Prentice Hall, New York, 1993.
4. F. Caccavale, M. Iamarino, F. Pierri, and V. Tufano. An adaptive controller-observer scheme for temperature control of non-chain reactions in batch reactors. *International Journal of Adaptive Control and Signal Processing*, 22:627–651, 2008.
5. F. Caccavale, F. Pierri, M. Iamarino, and V. Tufano. An integrated approach to fault diagnosis for a class of chemical batch processes. *Journal of Process Control*, 19:827–841, 2009.
6. C.C. Chang and C.C. Yu. On-line fault diagnosis using the signed directed graph. *Industrial and Engineering Chemistry Research*, 29(7):1290–1299, 1990.
7. C.T. Chang and J.W. Chen. Implementation issues concerning the EKF-based fault diagnosis techniques. *Chemical Engineering Science*, 50(18):2861–2882, 1995.
8. J. Chen and R.J. Patton. *Robust Model-Based Fault Diagnosis for Dynamic Systems*. Kluwer Academic, Dordrecht, 1999.

9. Y. Chetouani, N. Mouhab, J.M. Cosmao, and L. Estel. Application of extended Kalman filtering to chemical reactor fault detection. *Chemical Engineering Communications*, 189(9):1222–1241, 2002.
10. S.K. Dash, R. Rengaswamy, and V. Venkatasubramanian. Fault diagnosis in a nonlinear CSTR using observers. In *Proceedings of the 2001 Annual AIChE Meeting*, Reno, NV, p. 282i, 2001.
11. R. Dorr, F. Kratz, J. Ragot, F. Loisy, and J.L. Germain. Detection, isolation, and identification of sensor faults in nuclear power plants. *IEEE Transactions on Control Systems Technology*, 5(1):42–52, 1997.
12. R. Dunia and S. Joe Qin. Joint diagnosis of process and sensor faults using principal component analysis. *Control Engineering Practice*, 6:457–469, 1998.
13. P.M. Frank. Analytical and qualitative model-based fault diagnosis—a survey and some new results. *European Journal of Control*, 2:6–28, 1996.
14. P.M. Frank and X. Ding. Survey of robust residual generation and evaluation methods in observer-based fault detection systems. *Journal of Process Control*, 7:403–424, 1997.
15. J.B. Fussell. Fault tree analysis—state of the art. *IEEE Transactions on Reliability*, 23(1):51–53, 1974.
16. J. Gertler. Analytical redundancy methods in fault detection and diagnosis. In *Proceedings of IFAC SAFEPROCESS Symposium*, pages 9–21, 1991.
17. J.J. Gertler. *Fault Detection and Diagnosis in Engineering Systems*. Marcel Dekker, New York, 1998.
18. J. Gertler and D. Singer. A new structural framework for parity equation based failure detection and isolation. *Automatica*, 26:381–388, 1990.
19. D.M. Himmelblau. *Fault Detection and Diagnosis in Chemical and Petrochemical Processes*. Elsevier Press, Amsterdam, 1978.
20. J.C. Hoskins and D.M. Himmelblau. Artificial neural networks models of knowledge representation in chemical engineering. *Computers and Chemical Engineering*, 12:881–890, 1988.
21. Y. Huang, G.V. Reklaitis, and V. Venkatasubramanian. A heuristic extended Kalman filter based estimator for fault identification in a fluid catalytic cracking unit. *Industrial & Engineering Chemistry Research*, 42:3361–3371, 2003.
22. P.A. Ioannou and J. Sun. *Robust Adaptive Control*. Prentice Hall, Upper Saddle River, 1996.
23. R. Isermann. Process faults detection based on modelling and estimation methods—a survey. *Automatica*, 20(4):387–404, 1984.
24. P. Kaborè, S. Othman, T.F. McKenna, and H. Hammouri. Observer-based fault diagnosis for a class of nonlinear systems—application to a free radical copolymerization reaction. *International Journal of Control*, 73:787–803, 2000.
25. M. Karpenko, N. Sepehri, and D. Scuse. Diagnosis of process valve actuator faults using a multilayer neural network. *Control Engineering Practice*, 11:1289–1299, 2003.
26. P. Kesavan and J.H. Lee. A set based approach to detection of faults in multivariable systems. *Computers and Chemical Engineering*, 25:925–940, 2001.
27. H.K. Khalil. *Nonlinear Systems, 2nd Edition*. Prentice Hall, Upper Saddle River, 1996.
28. R. Li and J.H. Olson. Fault detection and diagnosis in a closed-loop nonlinear distillation process: application of extended Kalman filter. *Industrial Engineering Chemical Research*, 30(5):898–908, 1991.
29. J.F. MacGregor, J. Christiana, K. Costas, and M. Koutoudi. Process monitoring and diagnosis by multiblock PLS methods. *AIChE Journal*, 40(5):826–838, 1994.
30. R.S.H. Mah and A.C. Tamhane. Detection of gross errors in process data. *AIChE Journal*, 28:828, 1982.
31. A. Marciniak, C.D. Bocaniala, R. Louro, J. Sa da Costa, and J. Korbicz. Pattern recognition approach to fault diagnosis in the DAMADICS benchmark flow control valve. In *Proceedings of the 5th IFAC Symposium on Fault Detection, Supervision and Safety for Technical Processes*, pages 957–962. 2003.
32. N. Mehranbod, M. Soroush, M. Piovoso, and B.A. Ogunnaike. A probabilistic model for sensor fault detection and identification. *AIChE Journal*, 49(7):1787, 2003.
33. N. Mehranbod, M. Soroush, and C. Panjapornpon. A method of sensor fault detection and identification. *Journal of Process Control*, 15:321–339, 2005.

34. K. Patan and T. Parisini. Identification of neural dynamic models for fault detection and isolation: the case of a real sugar evaporation process. *Journal of Process Control*, 15:67–79, 2005.
35. R.J. Patton, P.M. Frank, and R.N. Clark. *Issues in Fault Diagnosis for Dynamic Systems*. Springer, London, 2000.
36. R.J. Patton, F.J. Uppal, and C.J. Lopez-Toribio. Soft computing approaches to fault diagnosis for dynamic systems: a survey. In *Preprints of the 4th IFAC Symposium on Fault Detection Supervision and Safety for Technical Processes*, Budapest, pages 298–311, 2001.
37. F. Pierri and G. Paviglianiti. Observer-based actuator fault detection for chemical batch reactors: a comparison between nonlinear adaptive and $\mathcal{H}_\infty$-based approaches. In *Proceedings of the Mediterranean Control Conference*, pages 1–6, 2007.
38. F. Pierri, G. Paviglianiti, F. Caccavale, and M. Mattei. Observer-based sensor fault detection and isolationfor chemical batch reactors. *Engineering Applications of Artificial Intelligence*, 21:1204–1216, 2008.
39. M.M. Polycarpou and A.J. Helmicki. Automated fault detection and accommodation: a learning systems approach. *IEEE Transactions on Systems, Man, and Cybernetics*, 25:1447–1458, 1995.
40. C. Rojas-Guzman and M.A. Kramer. Comparison of belief networks and rule-based expert systems for fault diagnosis of chemical processes. *Engineering Application of Artificial Intelligence*, 6:191, 1993.
41. D. Ruiz, J. Canton, J.M. Nougues, A. Espuña, and L. Puigjaner. On-line fault diagnosis system support for reactive scheduling in multipurpose batch chemical plants. *Computers and Chemical Engineering*, 25:829–837, 2001.
42. D. Ruiz, J.M. Nougues, and L. Puigjaner. Fault diagnosis support system for complex chemical plants. *Computers and Chemical Engineering*, 25:151–160, 2001.
43. N.J. Scenna. Some aspects of fault diagnosis in batch processes. *Reliability Engineering and System Safety*, 70(1):95–110, 2000.
44. O.A.Z. Sotomayor and D. Odloak. Observer-based fault diagnosis in chemical plants. *Chemical Engineering Journal*, 112:93–108, 2005.
45. R. Tarantino, F. Szigeti, and E. Colina-Morles. Generalized Luenberger observer-based fault detection filter design: An industrial application. *Control Engineering Practice*, 8:665–671, 2000.
46. H. Vedam and V. Venkatasubramanian. Signed digraph based multiple fault diagnosis. *Computers and Chemical Engineering*, 21:655–660, 1997.
47. H. Vedam and V. Venkatasubramanian. PCA-SDG based process monitoring and fault diagnosis. *Control Engineering Practice*, 7:903–917, 1999.
48. V. Venkatasubramanian, R. Vaidyanathan, and Y. Yamamoto. Process fault detection and diagnosis using neural networks—I steady state process. *Computers and Chemical Engineering*, 14:699–712, 1990.
49. V. Venkatasubramanian, R. Rengaswamy, and S.N. Kavuri. A review of process fault detection and diagnosis part II: Qualitative models and search strategies quantitative model-based methods. *Computers and Chemical Engineering*, 27:313–326, 2003.
50. V. Venkatasubramanian, R. Rengaswamy, K. Yin, and S.N. Kavuri. A review of process fault detection and diagnosis part I: quantitative model-based methods. *Computers and Chemical Engineering*, 27:293–311, 2003.
51. V. Venkatasubramanian, R. Rengaswamy, K. Yin, and S.N. Kavuri. A review of process fault detection and diagnosis part III: Process history based methods. *Computers and Chemical Engineering*, 27:327–346, 2003.
52. A.S. Willsky. A survey of design methods for failure detection in dynamic systems. *Automatica*, 12:601–611, 1976.
53. M. Witczak, J. Patton, and J. Korbicz. Fault detection with observers and genetic programming: application to the DAMADICS benchmark problem. In *Proceedings of the 5th IFAC Symposium on Fault Detection, Supervision and Safety for Technical Processes*, pages 1203–1208, 2003.

54. S. Yoon and J.F. MacGregor. Fault diagnosis with multivariate statistical models part I: using steady state fault signature. *Journal of Process Control*, 11:387–400, 2001.
55. D.L. Yu, J.B. Gomm, and D. Williams. Sensor fault diagnosis in a chemical process via RBF neural networks. *Control Engineering Practice*, 7:49–55, 1999.
56. X. Zhang, M.M. Polycarpou, and T. Parisini. A robust detection and isolation scheme for abrupt and incipient faults in nonlinear systems. *IEEE Transactions on Automatic Control*, 47:576–593, 2002.

Chapter 7
Applications to Nonideal Reactors

List of Principal Symbols

C	concentration [mol m^{-3}]
D	reactor diameter [m]
d_s	stirrer diameter [m]
f_{RT}	retention time distribution function
F_V	volumetric flow rate [m^3 s^{-1}]
Fr	dimensionless Froude number
g	gravitational acceleration [m s^{-2}]
h	heat transfer coefficient [J m^{-2} K^{-1} s^{-1}]
k	thermal conductivity [J m^{-1} K^{-1} s^{-1}]
K_P	power number
L	characteristic length [m]
N_c	number of reactor compartments
Nu	dimensionless Nusselt number
P	stirring power [J s^{-1}]
$\dot{Q}$	thermal power [J s^{-1}]
Re	dimensionless Reynolds number
S	heat exchange surface [m^2]
t	time [s]
t_b	batch time [s]
t_P	residence time [s]
U	overall heat transfer coefficient [J m^{-2} K^{-1} s^{-1}]
v	characteristic velocity [m s^{-1}]
V	volume [m^3]
V_c	compartment volume [m^3]

Greek Symbols

β	surface fraction
δ	scale-up ratio of linear dimensions
ν	kinematic viscosity [m^2 s^{-1}]
ρ	density [kg m^{-3}]

F. Caccavale et al., *Control and Monitoring of Chemical Batch Reactors*,
Advances in Industrial Control,
DOI 10.1007/978-0-85729-195-0_7,

χ dimensionless number defined in (7.6)
ω angular stirrer speed [rad s^{-1}]

Subscripts and Superscripts

av mean value
c main circulation flow
e secondary exchange flow
E exchanged
G gas
j jacket
ll lower level
r reactor
R reaction
S segregated conditions
ul upper level
us upstream

7.1 Nonideal Batch Reactors

In Chaps. 5 and 6 model-based control and early diagnosis of faults for ideal batch reactors have been considered. A detailed kinetic network and a correspondingly complex rate of heat production have been included in the mathematical model, in order to simulate a realistic application; however, the reactor was described by simple *ideal* mathematical models, as developed in Chap. 2. In fact, real chemical reactors differ from ideal ones because of two main causes of nonideal behavior, namely: the nonideal mixing of the reactor contents and the presence of multiphase systems.

Since real reactors are characterized by very different forms, it is deemed that the detailed analysis of a single real application may produce an increase in realism but also an unacceptable decrease in generality. Therefore, it is preferred to face these problems through theoretical analysis and mathematical modeling in order to give a general framework of the behavior of real batch reactors. This approach is hoped to be more useful both for developing practical applications of model-based control and fault diagnosis, and for suggesting future research work in this field.

In general, multiphase chemical reactors may contain any combination of different phases: solid, liquid, and gaseous. Figure 7.1 schematically shows a somewhat simplified classification of the batch reactors that are considered. In the simplest case (Fig. 7.1(a)), the reactor contains a liquid reacting mixture mixed by a mechanical stirrer, whereas the reactor head contains a gaseous phase made out of inert gases and/or of vapors, generated by the liquid phase components (solvent, reactants, products). In some cases, as shown in Fig. 7.1(b), the reactions occurring in the liquid phase produce a gaseous phase in the form of small bubbles that accumulates in the head and must be withdrawn from the reactor. In this respect, it is possible to distinguish the case of fermenters, in which the production of gas is usually accounted for by the kinetic model of the adopted reaction network, and the

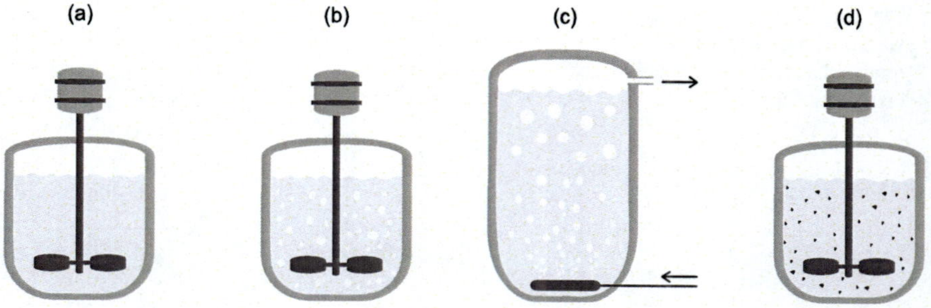

Fig. 7.1 Nonideal batch reactors: liquid-phase batch reactor (**a**), liquid-phase batch reactor with release of gaseous bubbles (**b**), semi-batch gas–liquid bubble column (**c**), and slurry batch reactor (**d**)

case of gassy reactions, in which the gases are produced by secondary unexpected side reactions.

Usually, the typology of batch reactors also includes the semi-batch gas–liquid reactors, in which a gaseous phase is fed continuously in order to provide one of the reactants. A typical example is given by the reactors used both in different oxidative industrial processes and in the active sludge processes for the treatment of wastewater. It is possible to distinguish between the bubble columns (Fig. 7.1(c)), in which the gas rises undisturbed in the liquid phase, and the bubble stirred reactor, in which a mechanical mixer is added. Finally, the slurry reactors can be considered, in which the liquid phase contains a finely dispersed solid phase as well, which can act as a reactant or as a heterogeneous catalyst; these reactors assume in general the features of Fig. 7.1(d).

Each of these four categories could be further subdivided when considering the operating temperature and pressure, the significance of thermal effects (exothermic or endothermic reactions) and the presence of heat exchange devices, the physical properties of the liquid phase (density, viscosity), the volume fraction occupied by the different phases in the reactors, and other minor effects. Notwithstanding these differences, once again, the modeling approach highlights the functional aspects, thus producing simpler and unifying descriptions.

7.2 Nonideal Mixing

As discussed in Sect. 2.1, physical and mathematical models of ideal chemical reactors are based on two very simplified fluid dynamic assumptions, namely perfect mixing (BR and CSTR) and perfect immiscibility (PFR). On the contrary, in real tank reactors the stirring system produces a complex motion field made out of vortices of different dimensions interacting with the reactor walls and the internal baffles, as schematically shown in Fig. 7.2(a). As a consequence, a complex field of composition and temperature is established inside the reactor.

The main features of this behavior may be captured by a simple modeling approach based on a proper combination of ideal reactors. The simplest example is

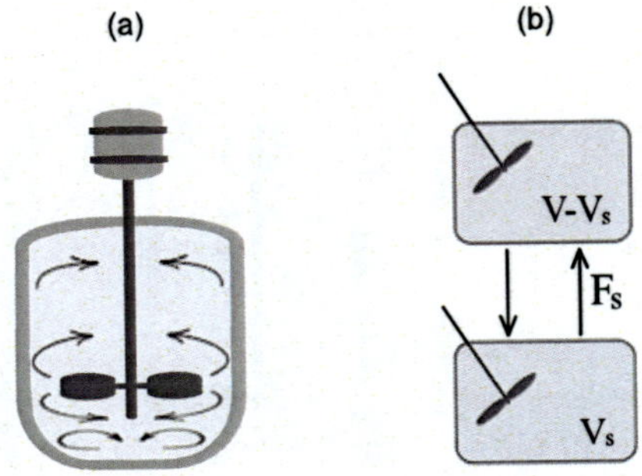

Fig. 7.2 Tank batch reactor with a segregated zone (**a**) modeled as two ideal suitably connected CSTRs (**b**)

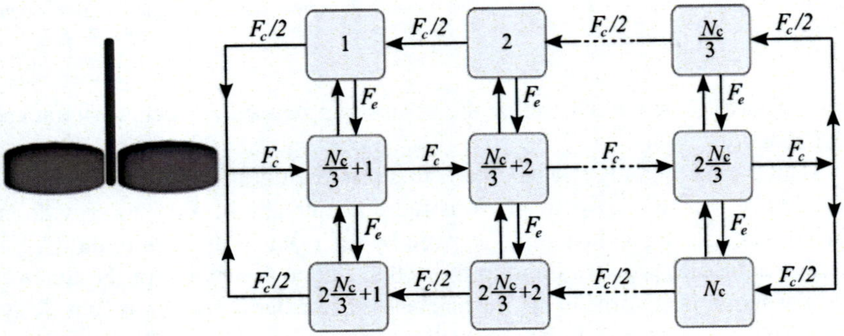

Fig. 7.3 Scheme of a compartment network around the impeller in a stirred tank reactor

the reactor with a segregated zone (Fig. 7.2(b)) described by two CSTRs suitably connected. In this model, two adjustable parameters are introduced, namely the segregated volume V_S and the relevant volumetric flow rate F_S. Since a very complex behavior is described by a very simple physical model and lumped in these parameters, their physical meaning is not straightforward, and the relevant values must be evaluated experimentally. In general, these parameters depend on the geometry of the reactor and the stirring system, on the rotation speed of the stirrer, and on the reactor dimensions; as discussed later in this chapter, the quality of mixing usually decreases when the dimensions of the reactor increase.

The modeling can be improved by increasing the number of ideal reactors so that a complex network composed of fictitious interacting compartments is created. This approach can be applied to describe a large number of industrial reactors, in which partial mixing phenomena take place [3, 10].

As an example, a compartment model for a single-phase jacketed batch reactor can be developed according to [4] when a vessel agitated by a Rushton turbine located halfway with respect to the liquid depth is considered. If the vorticity is eliminated by suitable baffles, the main liquid circulation flow rate F_c, generated by the impeller is radially directed and then split into two equal returning flows $F_c/2$, which are recirculated to the turbine (Fig. 7.3). Additionally, secondary exchange flow rates F_e must be considered to account for the axial mixing occurring at the ideal planes of separation between the main circulation streams. The circulation and exchange flow rates are related to both the impeller speed ω and the blade diameter d_s via the relationships [7, 9]

$$F_c = K_c \omega d_s^3 \tag{7.1}$$

and

$$F_e = K_e \omega d_s^3, \tag{7.2}$$

where the factors K_c and K_e mainly depend on the system geometry [4]. Hence, for any given system, a linear relationship between F_c and F_e holds.

On the basis of the considered macroscopic flow pattern, the dominant circulation flows (F_c and $F_c/2$) subdivide the reactor into three parallel levels, where each level is then divided into $N_c/3$ equally sized compartments of equal volume $V_c = V_r/N_c$. Every compartment is modeled as a nonstationary ideal continuous stirred tank reactor, with a main inlet and outlet flow, which connects the given compartment with adjacent compartments on the same level, and secondary exchange flow rates accounting for the turbulent mixing with adjacent compartments laying on the upper and/or lower level (Fig. 7.3).

The mathematical model of the reactor consists of the mass and energy balances written for all the compartments and an energy balance written for the jacket. The mass balance written for the reactant and for a first-order reaction in a generic compartment on the central level holds:

$$\dot{C} = -k_c C + \frac{F_c(C_{us} - C)}{V_c} + \frac{F_e(C_{ll} + C_{ul} - 2C)}{V_c}. \tag{7.3}$$

Here, subscripts u and l denote the concentrations in the adjacent compartment lying on the upper and lower levels, respectively, while subscript us denotes the concentration in the compartment located on the same level but upstream with respect to the dominant circulation flow. When this equation is written for the first compartment of the central level, the term accounting for the inlet circulation flow is modified, since it is composed by two different contributes at different concentrations originating from the corresponding compartments located on the upper and lower levels, respectively. Moreover, (7.3) must be suitably modified when the balance is referred to compartments laying on levels 1 and 3, where C_{ul} and C_{ll}, respectively, are equal to zero, and F_c has to be replaced by $F_c/2$.

The energy balance in the generic compartment on the central level yields

$$\dot{T} = \frac{(-\Delta H_R)k_c C}{\rho c_p} + \frac{\beta S U(T - T_j)}{V_c \rho c_p} + \frac{F_c(T_s - T)}{V_c} + \frac{F_e(T_{ll} + T_{ul} - 2T)}{V_c}, \tag{7.4}$$

where S is the total heat transfer area of the reactor, and β is the fraction of S facing the compartment considered ($\beta = 0$ for compartments not bordering the jacket). Moreover, the same corrections as for mass balances must be introduced in the energy balance written for the first compartment on the central level and for the compartments on levels 1 and 3.

Finally, the energy balance in the jacket reads

$$\dot{T}_\mathrm{j} = \sum_{i=1}^{N_\mathrm{j}} \frac{\beta_i SU(T_i - T_\mathrm{j})}{V_\mathrm{j}\rho_\mathrm{j}c_\mathrm{pj}} + \frac{(T_\mathrm{in} - T_\mathrm{j})}{V_\mathrm{j}} F_\mathrm{V}, \tag{7.5}$$

where the sum is extended to the N_j compartments bordering the jacket.

It has been shown [2] that, with reference to the mixing effectiveness, the reactor behavior can be described in terms of a dimensionless group χ, defined as the ratio of the length of the batch cycle t_b to the residence time $V_\mathrm{c}/F_\mathrm{c}$ of the fluid in each compartment,

$$\chi = \frac{F_\mathrm{c} t_\mathrm{b}}{V_\mathrm{c}}. \tag{7.6}$$

For low values of χ, noticeable temperature gradients may establish inside the reactor, with a consequent worsening of the controller performance. This effect depends on the sensor location as well. As an example, when temperature is measured in peripheral compartments, the higher temperatures established in the reactor core, i.e., in the proximity of the stirrer, are ignored. As a consequence, the average reaction rate and the rate of heat production are underestimated, so that the resulting control action is less effective in counteracting possible runaway phenomena.

In order to obtain a more detailed mathematical modeling, nonideal mixing must be considered in terms of Computational Fluid Dynamics (CFD). In CFD models the lumped parameter approach discussed above is substituted by a distributed parameter approach. By expanding the concept of CSTR, the reactor is described by a very large number of very small, perfectly mixed volumes, and the conservation equations are written in their differential form. In the simplest case (isothermal and nonreacting flow) the model is essentially made out of the Navier–Stokes equations and contains the tensor of turbulent stresses, which accounts for the turbulent transfer of momentum. This tensor depends not only on physical properties, such as density and laminar viscosity, but also on the geometry of the system and the whole motion field. In the general case, the corresponding turbulent terms for energy and mass transfer must be included, which show similar laws of dependence on physical properties, geometry, and fluid dynamics.

The description of small scale turbulent fields in confined spaces by fundamental approaches, based on statistical methods or on the concept of deterministic chaos, is a very promising and interesting research task; nevertheless, at the authors' knowledge, no fundamental approach is at the moment available for the modeling of large-scale confined systems, so that it is necessary to introduce semi-empirical models to express the tensor of turbulent stresses as a function of measurable quantities, such as geometry and velocity. Therefore, even in this case, a few parameters must be adjusted on the basis of independent measures of the fluid dynamic behavior. In any case, it must be underlined that these models are very complex and, therefore, well suited for simulation of complex systems but neither for identification of chemical parameters nor for online control and diagnosis [5, 6].

7.3 Multiphase Batch Reactors

The main effect of the presence of gases and vapors in the reactor head is the relevance of an additional variable, the pressure. In a simple mathematical model, each phase may be described as a well-mixed volume (even if no mechanical stirrer is present in the gaseous phase), and it is also possible to consider thermal and mechanical equilibrium between the phases, i.e., to set equal values of temperature and pressure in the two phases.

A few additional equations must be included in the mathematical model, namely a mass balance and an equation of state for the gaseous phase, and the equilibrium relationships between the phases, i.e., a distinct equation for each component present in both phases. Moreover, the heat balance for the reacting mixture must be modified by including the heat of evaporation, which usually has a nonnegligible effect on the reactor temperature. This augmented mathematical model is complex enough to hinder a unified approach; an example may be found in [8], where some guidelines are given for the design of suppression systems against chemical runaway.

In fact, since vapor pressure is usually a strong positive function of temperature, pressure is the most sensible variable to evaluate the possible occurrence of an explosion, which is just defined as an uncontrolled growth of pressure. A similar approach allows considering the more dangerous case of runaway represented by the so-called gassy reactions, i.e., the case in which gaseous products are generated by the chemical reaction, irrespective of thermal effects. In the case of gassy reaction, an increase of pressure can be observed even in the absence of a significant increase of temperature, thus presenting a severe challenge for the reactor controller, since pressure sensors are usually less sensible and less accurate than temperature sensors, as discussed in Sect. 2.6.

A more complex behavior is expected when multiphase reacting systems are examined. As an example, consider the gas–liquid reactor sketched in Fig. 7.1(c), which behaves as a batch reactor with respect to the liquid phase and as a continuous reactor with respect to the gaseous phase. A reactant is transferred from the gaseous to the liquid phase, where it reacts with a substrate.

In the absence of a stirring system, the bubbles rise undisturbed through the liquid phase at a velocity that depends on their dimensions. This bubble phase can be modeled as a plug flow reactor, whereas the liquid phase can be described by one or a few well-mixed reactors in series (Fig. 7.4(a)). On the contrary, when a mechanical stirrer is included in the liquid phase, a well-mixed model can be also considered for the bubble phase (Fig. 7.4(b)). In fact, the stirrer captures the bubbles in the liquid vortices, thus increasing their average residence time in the liquid phase, and, moreover, it also produces a fragmentation and coalescence mechanism which mixes the contents of the gaseous bubbles. The PFR model for the gaseous phase can be more appropriate than the CSTR model not only when the stirrer speed is very low but also whenever a preferential bubble channel is formed along the stirrer axis, as it is sometimes observed for very large stirrer speeds.

The mathematical models corresponding to the schemes in Fig. 7.4 contain the diffusive fluxes between the phases, which depend both on fluid dynamics (through

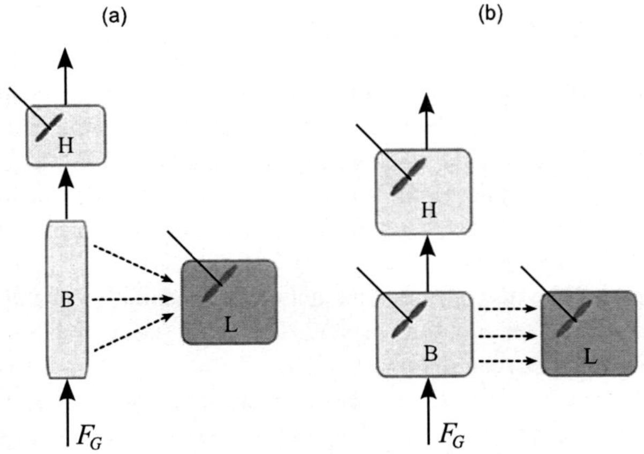

Fig. 7.4 Schemes for mathematical models of a gas–liquid bubble column (**a**) and a gas–liquid stirred reactor (**b**). B = bubble phase, H = reactor head, L = liquid phase, F_G = gas flow rate; *dashed arrows* represent diffusive interphase flux

a suitable mass transfer coefficient and on the interfacial area between the phases) and the difference between the concentrations of the transferred component in the two phases. The diffusive fluxes connect different phases present in the reactor, thus playing a role similar to that of the volumetric flow rate F_S connecting segregated volumes in partially mixed monophasic reactors. Nevertheless, in practical cases, this segregation effect between phases can be quantitatively more noticeable than the simple fluid dynamic segregation of different volumes within a single phase. Of course, in the presence of noticeable thermal effects, interphase heat transfer may be observed as well.

7.4 Scaling-up the Information

The performance of real chemical reactors depends on the interaction between chemical kinetics, fluid dynamics, and interphase transport phenomena. Therefore, it is not surprising that two geometrically similar reactors (i.e., having equal shape) usually show a behavior strongly dependent on dimensions. Therefore, the design procedures of real chemical reactors make a wide use of the concept of scaling up to the industrial scale the information obtained at the laboratory and/or at the pilot scale. Since control and fault diagnosis performance is affected by reactor dimensions, a brief account of these concepts may be useful in the present context in order to obtain a deeper understanding of the behavior of real chemical reactors.

7.4.1 Basic Ideas of Scale-up

Scaling is a very important tool in science and engineering, because, in general, the physical behavior of a system is not independent from dimensions. The first approach to scaling consists in writing the physical laws in nondimensional form in order to derive the expressions for the relevant dimensionless groups; in fact, if these groups are kept constant on increasing dimensions, the dimensionless solutions are simply scaled as functions of space and time. Perhaps, the simplest example is the fluid dynamic behavior of fluids in smooth tubes, which is scaled by simply taking as a constant the ratio between inertia and drag forces expressed by the Reynolds number $Re = vL/\nu$, where L is a characteristic length, v is a characteristic velocity, and ν is the kinematic viscosity.

Unfortunately, real things are rarely simple: a more complex example from fluid dynamics is taken from [1] and refers to the vortex depth in unbaffled agitated tanks. The dimensionless equation of motion shows that the dynamic similarity may be obtained in two geometrically similar tanks of different diameter ($D_2 > D_1$), when the same values are fixed for the two relevant dimensionless numbers, namely the Reynolds number and the Froude number, which represents the ratio between inertia and gravity forces. Here, the impeller diameter d_s is taken as the reference length, whereas the reference velocity is taken to be proportional to the velocity of the stirrer tips, i.e., the product between d_s and the impeller angular speed ω; thus, the Reynolds number is defined as $Re = \omega d_s^2/\nu$, and the Froude number as $Fr = \omega^2 d_s^2/g d_s = \omega^2 d_s/g$, where g denotes the gravitational acceleration.

Since two geometrically similar tanks are considered, the ratio $\delta = d_{s2}/d_{s1}$ equals the ratio D_2/D_1 of the reactor diameters, and the vortex depths are in the same ratio if $Re_2 = Re_1$ and $Fr_2 = Fr_1$. It can be shown that, at constant ν, these two conditions of dynamic similarity, respectively, yield the following scale-up prescriptions:

$$\frac{\omega_2}{\omega_1} = \left(\frac{D_1}{D_2}\right)^2 = \delta^2 \tag{7.7}$$

and

$$\frac{\omega_2}{\omega_1} = \sqrt{\frac{D_1}{D_2}} = \sqrt{\delta}, \tag{7.8}$$

which cannot be satisfied at the same time if the same fluid is used in the two tanks. In conclusion, a proportionally deeper vortex is obtained in the smaller tank, whereas it is possible to restore dynamic similarity by using in the smaller tank a fluid of smaller kinematic viscosity. Of course, this last condition, which is widely used in physical modeling of large natural systems, cannot be satisfied in chemical reactors.

A second example of interest in the present context refers to the scaling of thermal effects. Any object (a chemical reactor such as a living body) that produces heat at a rate proportional to its volume ($\dot{Q}_R \propto V_r$) and exchanges heat with a cooling device or with the ambient at a rate proportional to its lateral surface S_L and to the temperature difference with respect to the external heat sink (i.e., $\dot{Q}_E = U S_L (T_r - T_a)$) can maintain the same temperature, independently of its dimensions, only if the ratio $U S_L / V_r$ is kept constant. In general, this condition cannot be satisfied, since the ratio S_L / V_r is inversely proportional to the characteristic linear dimension, and the overall heat exchange coefficient U is a complex function of dimensions and of fluid dynamics.

7.4.2 The Scale-up of Real Batch Reactors

The principles and methods of scale-up can be applied to chemical reactors. In the absence of significant thermal effects, i.e., when the ratio $\dot{Q}_R/V_r$ may be considered negligible, ideal batch reactors do not show any problem of scale-up, because the volume V_r does not appear in the mathematical model (2.17), so that their performance is only determined by chemical kinetics (see Sect. 2.3). On the contrary, a very complex behavior is expected for real reactors; in fact, this behavior cannot be analyzed in terms of mathematical models, and the design procedures must be largely based on semi-empirical rules of scale-up.

For the sake of completeness, two different scale-up rules are briefly considered here, and their different results are shown. First, let us define the stirring power P as

$$P = K_P \rho \omega^3 d_s^5, \tag{7.9}$$

where the dimensionless power number K_P represents a drag coefficient which depends on the geometry of the stirrer and on the Reynolds number, again defined as $Re = \omega d_s^2/\nu$. If both (geometrically similar) tank reactors are operated at elevated Reynolds numbers, it is possible to assume that $N_{P2} = N_{P1}$, so that (7.9) yields

$$\frac{P_2}{P_1} = \left(\frac{\omega_2}{\omega_1}\right)^3 \left(\frac{d_{s2}}{d_{s1}}\right)^5 = \left(\frac{\omega_2}{\omega_1}\right)^3 \delta^5. \tag{7.10}$$

If the power per unit volume P/V_r is assumed to determine the effectiveness of the mixing, it is possible to introduce the scale-up criterion $P/V_r = \text{constant}$, which gives the scale-up prescriptions

$$\frac{P_2}{P_1} = \delta^2 \tag{7.11}$$

and

$$\frac{\omega_2}{\omega_1} = \delta^{-2/3}. \tag{7.12}$$

An alternative criterion is derived by considering that the velocity of the stirrer tips, $v = \pi d_s \omega$, determines the shear stress on the fluid and, consequently, its micro mixing. The scale-up criterion $d_s \omega = \text{constant}$ gives

$$\frac{P_2}{P_1} = \delta^3 \tag{7.13}$$

and

$$\frac{\omega_2}{\omega_1} = \delta^{-1}, \tag{7.14}$$

so that, on increasing the reactor volume, the required stirrer speed decreases more strongly than in the previous case; correspondingly, the required stirring power increases less strongly.

In the presence of significant thermal effects, one of the above-mentioned fluid dynamic scale-up criteria must be considered, together with the criterion $US/V =$ constant, which can be made more realistic by considering that heat exchange surface S can be different from S_L and by introducing a proper functional relationship for U. If the internal resistance to heat transfer prevails, U may be intended as the internal heat transfer coefficient h, so that the relationship

$$h \propto \omega^{0.65} d_s^{1.3} D^{-1} \propto \omega^{0.65} d_s^{0.3} \tag{7.15}$$

holds, as it can be easily derived from a dimensionless correlation of the type $Nu = (hD)/k \propto Re^{0.65}$.

In general, on increasing the reactor volume, the required heat transfer surface increases faster than δ^3; this effect is enhanced by the decrease of the heat transfer coefficient. This prescription cannot be obeyed by the lateral surface of the reactor (which increases as δ^2) so that an internal or external additional heat exchange surface, whose dimensions can be fixed independently from the reactor dimensions, must be provided.

7.5 Suggestions and Conclusions

It is hoped that the arguments developed in this chapter can provide some useful suggestions both for developing practical applications of the control and diagnosis approaches developed in this book to real batch reactors and for planning new research work in this field. No definitive conclusion of general validity is expected from these activities because of the very different forms assumed by real reactors; nevertheless, some expected results can be anticipated when considering a few simple cases.

The case of gassy reactions represents a severe test for any control system. Since in this case a pressure increase is almost independent of the temperature increase, the system must be monitored by means of pressure sensors, and these are, in general, less accurate and reliable when compared to temperature sensors. In the more general case of nonideal mixing in homogeneous reacting media, it appears that the effect of segregated zones may be that of transforming the reactor in a higher-order system. Thus, not only the response of the system is slowed down, but, in some cases, an oscillating behavior may be observed.

This effect can be forecast on the basis of the retention time distribution function in continuous tank reactors, which represents the simplest approach to the analysis of reactor dynamics. In its cumulative form, this function represents, for any time t, the fraction of the exit volumetric flow rate characterized by a residence time smaller than t and can be measured experimentally by submitting the reactor to a step forcing input in the entering stream. Whereas for the ideal tank reactor, the following simple exponential function is obtained:

$$f_{\mathrm{RT}}(t) = 1 - \exp\left(-\frac{t}{t_{\mathrm{P,av}}}\right), \tag{7.16}$$

in which the ratio $t_{\mathrm{P,av}} = V_{\mathrm{r}}/F_{\mathrm{V}}$ between volume and volumetric flow rate represents the average residence time, in the presence of segregated volumes more complex functions, which present a zero derivative at $t = 0$ and inflection points, are obtained.

A similar, but more significant, effect is expected for multiphase reactors. A general conclusion may be that the more the reactors are complex, the more advantageous is the use of model-based approaches, when compared to more empirical ones. This is true from a practical point of view as well, since the increasing availability of fast and low-price computing devices allows improving the complexity of the models; it is deemed that the limits to this approach depend essentially on the quality of the experimental data available for identification purposes.

References

1. R.B. Bird, W.E. Stewart, and E.N. Lightfoot. *Transport Phenomena, 2nd Edition*. Wiley, New York, 2001.
2. F. Caccavale, M. Iamarino, F. Pierri, G. Satriano, and V. Tufano. Effect of non-ideal mixing on control of cooled batch reactors. In *Proc. of 5th Mathmod*, Wien, 2006.
3. S. Claudel, C. Fonteix, J.P. Leclerc, and H.G. Lintz. Application of the possibility theory to the compartment modelling of flow pattern in industrial processes. *Chemical Engineering Science*, 58:4005–4016, 2003.
4. Y.Q. Cui, R.G.J.M. Van der Lans, H.J. Noorman, and K.C.A.M. Luyben. Compartment mixing model for stirred reactors with multiple impellers. *Transactions of IChemE*, 74:261–271, 1996.
5. B.E. Launder and D.B. Spalding. *Lectures in Mathematical Models for Turbulence*. Academic Press, San Diego, 1972.
6. M. Lesieur. *Turbulence in Fluids, 4th Edition*. Springer, Dordrecht, 2008.
7. A.W. Nienow. On impeller circulation and mixing effectiveness in the turbulent flow regime. *Chemical Engineering Science*, 52:2557–2565, 1997.
8. V. Tufano. Modeling runaway reactions in reactors protected with suppression systems. *Journal of Hazardous Materials*, 19:225–236, 1988.
9. J.M.T. Vasconcelos, S.S. Alves, and J.M. Barata. Mixing in gas–liquid contactors agitated by multiple turbines. *Chemical Engineering Science*, 50:2343–2354, 1995.
10. J. Zhang and R. Smith. Design and optimisation of batch and semi-batch reactors. *Chemical Engineering Science*, 59:459–478, 2004.

Appendix A
Proofs

A.1 Proof of Theorem 5.1

The error dynamics can be readily derived from (5.15) and (5.23):

$$\begin{cases} \dot{\tilde{\boldsymbol{x}}} = \boldsymbol{A}_{\rm o}(y)\tilde{\boldsymbol{x}} + \boldsymbol{C}^{\rm T}\boldsymbol{\psi}(y)\tilde{\theta}_{\rm o}, \\ \tilde{y} = \boldsymbol{C}\tilde{\boldsymbol{x}}, \end{cases} \tag{A.1}$$

where $\boldsymbol{A}_{\rm o}(y) = \boldsymbol{A}(y) - \boldsymbol{L}\boldsymbol{C}$. Let us consider the following positive definite Lyapunov candidate function:

$$V_{\rm o}(\tilde{\boldsymbol{x}}, \tilde{\theta}_{\rm o}) = \frac{1}{2}\tilde{\boldsymbol{x}}^{\rm T}\boldsymbol{P}_{\rm o}\tilde{\boldsymbol{x}} + \frac{1}{2}\gamma_{\rm o}\tilde{\theta}_{\rm o}^2, \tag{A.2}$$

where $\boldsymbol{P}_{\rm o}$ is the positive definite diagonal matrix

$$\boldsymbol{P}_{\rm o} = \mathrm{diag}\{\sigma_1, \ldots, \sigma_{N_{\rm C}}, 1, 1\}, \tag{A.3}$$

and the σ_is are constant positive values to be determined.

The derivative of $V_{\rm o}$ along the trajectories of the error dynamics, taking into account that the parameter θ is constant, is given by

$$\begin{aligned} \dot{V}_{\rm o} = &-\sum_{i=1}^{N_{\rm C}} \sigma_i k_{{\rm c}i}\tilde{x}_i^2 - l_{\rm r}\tilde{x}_{N_{\rm C}+1}^2 - l_{\rm j}\tilde{x}_{N_{\rm C}+2}^2 + \sum_{i=1}^{N_{\rm C}-1}\sum_{h=i+1}^{N_{\rm C}} \sigma_h \upsilon_{i,h} k_{{\rm c}i,h}\tilde{x}_i\tilde{x}_h \\ &+ \sum_{i=1}^{N_{\rm C}} (a_i - \sigma_i l_i)\tilde{x}_i\tilde{x}_{N_{\rm C}+1} + \boldsymbol{\psi}^{\rm T}(y)\boldsymbol{C}\tilde{\boldsymbol{x}}\tilde{\theta}_{\rm o} - \gamma_{\rm o}\dot{\hat{\theta}}_{\rm o}\tilde{\theta}_{\rm o}, \end{aligned}$$

where the dependence of the rate constants upon the temperature has been dropped for notation compactness. By considering the update law (5.24) and inequali-

F. Caccavale et al., *Control and Monitoring of Chemical Batch Reactors*,
Advances in Industrial Control,
DOI 10.1007/978-0-85729-195-0,

ties (2.32) and (2.33), $\dot{V}_\mathrm{o}$ can be bounded as follows:

$$\dot{V}_\mathrm{o} \le -\sum_{i=1}^{N_\mathrm{C}} \sigma_i \underline{k}_{\mathrm{c}i} \widetilde{x}_i^2 - l_\mathrm{r} \widetilde{x}_{N_\mathrm{C}+1}^2 - l_\mathrm{j} \widetilde{x}_{N_\mathrm{C}+2}^2$$

$$+ \sum_{i=1}^{N_\mathrm{C}-1} \sum_{h=i+1}^{N_\mathrm{C}} \sigma_h \upsilon_{i,h} \overline{k}_{\mathrm{c}i,h} |\widetilde{x}_i||\widetilde{x}_h| + \sum_{i=1}^{N_\mathrm{C}} (\overline{a}_i + \sigma_i l_i)|\widetilde{x}_i||\widetilde{x}_{N_\mathrm{C}+1}|$$

$$= -\sum_{i=1}^{N_\mathrm{C}-1} \sum_{h=i+1}^{N_\mathrm{C}} \begin{bmatrix} |\widetilde{x}_i| \\ |\widetilde{x}_h| \end{bmatrix}^\mathrm{T} \boldsymbol{\Omega}_{i,h} \begin{bmatrix} |\widetilde{x}_i| \\ |\widetilde{x}_h| \end{bmatrix} - \sum_{i=1}^{N_\mathrm{C}} \begin{bmatrix} |\widetilde{x}_i| \\ |\widetilde{x}_{N_\mathrm{C}+1}| \end{bmatrix}^\mathrm{T} \boldsymbol{\Phi}_i \begin{bmatrix} |\widetilde{x}_i| \\ |\widetilde{x}_{N_\mathrm{C}+1}| \end{bmatrix} - l_\mathrm{j} \widetilde{x}_{N_\mathrm{C}+2}^2,$$

where

$$\overline{a}_i = \sum_{h=i+1}^{N_\mathrm{C}+1} \alpha_{i,h} \overline{k}_{\mathrm{c}i,h}. \tag{A.4}$$

The matrices on the right-hand side of the above inequality,

$$\boldsymbol{\Omega}_{i,h} = \begin{bmatrix} \frac{\sigma_i \underline{k}_{\mathrm{c}i}}{N_\mathrm{C}} & -\frac{\sigma_h \upsilon_{i,h} \overline{k}_{\mathrm{c}i,h}}{2} \\ -\frac{\sigma_h \upsilon_{i,h} \overline{k}_{\mathrm{c}i,h}}{2} & \frac{\sigma_h \underline{k}_{\mathrm{c}h}}{N_\mathrm{C}} \end{bmatrix}, \qquad \boldsymbol{\Phi}_i = \begin{bmatrix} \frac{\sigma_i \underline{k}_{\mathrm{c}i}}{N_\mathrm{C}} & -\frac{\overline{a}_i + \sigma_i l_i}{2} \\ -\frac{\overline{a}_i + \sigma_i l_i}{2} & \frac{l_\mathrm{r}}{N_\mathrm{C}} \end{bmatrix}, \tag{A.5}$$

are all positive definite if the gains satisfy the inequality

$$l_\mathrm{r} > \max_{i=1,\dots,N_\mathrm{C}} \left\{ \frac{N_\mathrm{C}^2 (\overline{a}_i + \sigma_i l_i)^2}{4 \sigma_i \underline{k}_{\mathrm{c}i}} \right\} \tag{A.6}$$

and the positive constants σ_is satisfy the inequalities

$$\sigma_i > \max_{h=i+1,\dots,N_\mathrm{C}} \left\{ \frac{N_\mathrm{C}^2 \upsilon_{i,h}^2 \overline{k}_{\mathrm{c}i,h}^2}{4 \underline{k}_{\mathrm{c}i} \underline{k}_{\mathrm{c}h}} \sigma_h \right\}, \quad i = N_\mathrm{C} - 1, \dots, 1. \tag{A.7}$$

Therefore, $\dot{V}_\mathrm{o}$ can be upper bounded as follows:

$$\dot{V}_\mathrm{o} \le -\sum_{i=1}^{N_\mathrm{C}-1} \sum_{h=i+1}^{N_\mathrm{C}} \lambda_{\min}(\boldsymbol{\Omega}_{i,h}) \big(\widetilde{x}_i^2 + \widetilde{x}_h^2 \big)$$

$$- \sum_{i=1}^{N_\mathrm{C}} \lambda_{\min}(\boldsymbol{\Phi}_i) \big(\widetilde{x}_i^2 + \widetilde{x}_{N_\mathrm{C}+1}^2 \big) - l_\mathrm{j} \widetilde{x}_{N_\mathrm{C}+2}^2, \tag{A.8}$$

where $\lambda_{\min}(\cdot)$ is the smallest eigenvalue of a matrix. Thus,

$$\dot{V}_\mathrm{o} \le -\zeta_\mathrm{o} \|\widetilde{\boldsymbol{x}}\|^2, \tag{A.9}$$

where

$$\zeta_o = \min\{(N_C - 1)\underline{\omega} + \underline{\phi}, N_C\underline{\phi}, l_j\},$$

$$\underline{\omega} = \min_{\substack{i=1,\dots,N_C-1\\ h=i+1,\dots,N_C}} \{\lambda_{\min}(\boldsymbol{\Omega}_{i,h})\}, \qquad \underline{\phi} = \min_{i=1,\dots,N_C} \{\lambda_{\min}(\boldsymbol{\Phi}_i)\}.$$

Hence, $\dot{V}_o$ is negative semi-definite: this guarantees the boundedness of $\widetilde{\boldsymbol{x}}$ and $\widetilde{\theta}_o$. By invoking Barbalat's lemma [3], it can be recognized that $\dot{V}_o \to 0$, which implies the global uniform convergence to $\boldsymbol{0}$ of $\widetilde{\boldsymbol{x}}$ as $t \to \infty$, while $\widetilde{\theta}_o$ is only guaranteed to be uniformly bounded. □

A.2 Proof of Theorem 5.2

The estimation error dynamics can be derived from (5.27) and (5.29):

$$\begin{cases} \dot{\widetilde{\boldsymbol{x}}}_E = \boldsymbol{A}_{E,o}(\theta)\widetilde{\boldsymbol{x}}_E + \boldsymbol{\psi}(\widetilde{\boldsymbol{y}})\widetilde{\theta}_o + \widetilde{\boldsymbol{\xi}}(\boldsymbol{y}, \widetilde{\boldsymbol{\theta}}_q), \\ \widetilde{\boldsymbol{y}} = \widetilde{\boldsymbol{x}}_E, \end{cases} \tag{A.10}$$

where $\boldsymbol{A}_{E,o} = \boldsymbol{A}_E - \boldsymbol{L}_E$ and

$$\widetilde{\boldsymbol{\xi}}(\boldsymbol{y}, \widetilde{\boldsymbol{\theta}}_q) = \begin{bmatrix} \widetilde{\boldsymbol{\theta}}_q^T \boldsymbol{\varphi}(\boldsymbol{y}) \\ 0 \end{bmatrix}.$$

Let us consider the following positive definite Lyapunov candidate function:

$$V_o(\widetilde{\boldsymbol{x}}_E, \widetilde{\theta}_o, \widetilde{\boldsymbol{\theta}}_q) = \frac{1}{2}\widetilde{\boldsymbol{x}}_E^T\widetilde{\boldsymbol{x}}_E + \frac{1}{2}\gamma_o\widetilde{\theta}_o^2 + \frac{1}{2}\gamma_q\widetilde{\boldsymbol{\theta}}_q^T\widetilde{\boldsymbol{\theta}}_q. \tag{A.11}$$

The derivative of V_o along the trajectories of the error dynamics, taking into account that the parameters θ and $\boldsymbol{\theta}_q$ are constant, is given by

$$\begin{aligned} \dot{V}_o = &-(\alpha_r\theta + l_r)\widetilde{x}_{N_C+1}^2 - (\alpha_j\theta + l_j)\widetilde{x}_{N_C+2}^2 + (\alpha_r\theta + \alpha_j\theta)\widetilde{x}_{N_C+1}\widetilde{x}_{N_C+2} \\ &+ \boldsymbol{\psi}(\widetilde{\boldsymbol{y}})^T\widetilde{\boldsymbol{y}}\widetilde{\theta}_o - \gamma_o\widetilde{\theta}_o\dot{\widehat{\theta}}_o + \widetilde{\boldsymbol{\theta}}_q^T\boldsymbol{\varphi}(\boldsymbol{y})\widetilde{y}_1 - \gamma_q\widetilde{\boldsymbol{\theta}}_q^T\dot{\widehat{\boldsymbol{\theta}}}_q. \end{aligned}$$

By considering the update laws (5.24) and (5.34), $\dot{V}_o$ can be bounded as follows:

$$\begin{aligned} \dot{V}_o &\le -(\alpha_r\theta + l_r)\widetilde{x}_{N_C+1}^2 - (\alpha_j\theta + l_j)\widetilde{x}_{N_C+2}^2 + (\alpha_r\theta + \alpha_j\theta)|\widetilde{x}_{N_C+1}||\widetilde{x}_{N_C+2}| \\ &= -\begin{bmatrix} |\widetilde{x}_{N_C+1}| \\ |\widetilde{x}_{N_C+2}| \end{bmatrix}^T \begin{bmatrix} \alpha_r\theta + l_r & -\frac{\alpha_r+\alpha_j}{2}\theta \\ -\frac{\alpha_r+\alpha_j}{2}\theta & \alpha_j\theta + l_j \end{bmatrix} \begin{bmatrix} |\widetilde{x}_{N_C+1}| \\ |\widetilde{x}_{N_C+2}| \end{bmatrix}. \end{aligned} \tag{A.12}$$

The matrix on the right hand-side of the above inequality is positive definite if the gains satisfy the following inequality:

$$l_j > \frac{(\alpha_r + \alpha_j)^2\theta^2}{4(\alpha_r\theta + l_r)} - \alpha_j\theta. \tag{A.13}$$

Therefore, $\dot{V}_o$ can be upper bounded as follows:

$$\dot{V}_o \leq -\zeta_o \|\widetilde{\boldsymbol{x}}_E\|^2, \tag{A.14}$$

where ζ_o is the minimum eigenvalue of the matrix in (A.12).

Hence, $\dot{V}_o$ is negative semi-definite: this guarantees the boundedness of $\widetilde{\boldsymbol{x}}$, $\widetilde{\theta}_o$, and $\widetilde{\boldsymbol{\theta}}_q$. By invoking Barbalat's lemma [3], it can be recognized that $\dot{V}_o \to 0$, which implies the global uniform convergence to $\mathbf{0}$ of $\widetilde{\boldsymbol{x}}$ as $t \to \infty$, while $\widetilde{\theta}_o$ and $\widetilde{\boldsymbol{\theta}}_q$ are only guaranteed to be uniformly bounded. □

A.3 Proof of Theorem 5.3

The closed-loop dynamics can be derived by plugging (5.38) and (5.39) into (5.15) and taking into account that $y_2 = y_{2,\mathrm{des}} - e_2$ and $\widetilde{\theta}_c = \theta - \widehat{\theta}_c$ (and thus, $\theta = \widehat{\theta}_c + \widetilde{\theta}_c$):

$$\dot{\boldsymbol{\varepsilon}} = \boldsymbol{A}_c \boldsymbol{\varepsilon} + \boldsymbol{\chi}_c(\boldsymbol{y})\widetilde{\theta}_c - \boldsymbol{A}_{co}(\boldsymbol{y})\widetilde{\boldsymbol{x}}, \tag{A.15}$$

where the matrices $\boldsymbol{A}_c$ and $\boldsymbol{A}_{co}$ are given by

$$\boldsymbol{A}_c = \begin{bmatrix} \boldsymbol{A}_r & \boldsymbol{A}_{rj} \\ \boldsymbol{O}_{2\times 2} & \boldsymbol{A}_j \end{bmatrix}, \qquad \boldsymbol{A}_{co}(\boldsymbol{y}) = \begin{bmatrix} \mathbf{0}_{1\times N_C} & \mathbf{0}_{1\times 2} \\ \boldsymbol{a}^{\mathrm{T}}(\boldsymbol{y}) & \mathbf{0}_{1\times 2} \\ \boldsymbol{O}_{2\times N_C} & \boldsymbol{O}_{2\times 2} \end{bmatrix}.$$

Let us consider the Lyapunov candidate function

$$V(\widetilde{\boldsymbol{x}}, \boldsymbol{\varepsilon}, \widetilde{\theta}_o, \widetilde{\theta}_c) = V_o(\widetilde{\boldsymbol{x}}, \widetilde{\theta}_o) + \varrho V_c(\boldsymbol{\varepsilon}, \widetilde{\theta}_c), \tag{A.16}$$

where V_o is the function defined in (A.2), V_c is given by

$$V_c(\boldsymbol{\varepsilon}, \widetilde{\theta}_c) = \frac{1}{2}\boldsymbol{\varepsilon}^{\mathrm{T}} \boldsymbol{P}_c \boldsymbol{\varepsilon} + \frac{1}{2}\gamma_c \widetilde{\theta}_c^2, \tag{A.17}$$

and $\varrho > 0$ is a positive constant to be determined.

The derivative of V_c along the trajectories of the system (A.1), (A.15), taking into account that the parameter θ is constant, is given by

$$\dot{V}_c = -\frac{1}{2}\boldsymbol{\varepsilon}^{\mathrm{T}} \boldsymbol{N}_c \boldsymbol{\varepsilon} + \boldsymbol{\chi}_c^{\mathrm{T}}(\boldsymbol{y}) \boldsymbol{P}_c \boldsymbol{\varepsilon} \widetilde{\theta}_c - \gamma_c \dot{\widehat{\theta}}_c \widetilde{\theta}_c - \boldsymbol{\varepsilon}^{\mathrm{T}} \boldsymbol{P}_c \boldsymbol{A}_{co}(\boldsymbol{y})\widetilde{\boldsymbol{x}}, \tag{A.18}$$

where

$$\boldsymbol{N}_c = \begin{bmatrix} \boldsymbol{N}_r & -\boldsymbol{P}_r \boldsymbol{A}_{rj} \\ -\boldsymbol{A}_{rj}^{\mathrm{T}} \boldsymbol{P}_r & \boldsymbol{N}_j \end{bmatrix}$$

is a symmetric matrix, which turns out to be positive definite if $\boldsymbol{N}_r$ and $\boldsymbol{N}_j$ satisfy (5.44). Finally, taking into account the update law (5.40), $\dot{V}_c$ becomes

$$\dot{V}_c = -\frac{1}{2}\boldsymbol{\varepsilon}^{\mathrm{T}} \boldsymbol{N}_c \boldsymbol{\varepsilon} - \boldsymbol{\varepsilon}^{\mathrm{T}} \boldsymbol{P}_c \boldsymbol{A}_{co}(\boldsymbol{y})\widetilde{\boldsymbol{x}}. \tag{A.19}$$

Hence, $\dot{V}_c$ can be upper bounded as follows:

$$\dot{V}_c \leq -\zeta_c \|\boldsymbol{\varepsilon}\|^2 + \zeta_{co} \|\widetilde{\boldsymbol{x}}\| \|\boldsymbol{\varepsilon}\|, \tag{A.20}$$

where $\zeta_c = \lambda_{\min}(\boldsymbol{N}_c)/2$ and

$$\zeta_{co} = \|\boldsymbol{P}_c\| \max_{\boldsymbol{y}} \{ \|\boldsymbol{A}_{co}(\boldsymbol{y})\| \},$$

since (2.33) ensures that $\boldsymbol{A}_{co}$ is norm-bounded for any $\boldsymbol{y}$.

Therefore, by considering the inequalities in (A.9) and (A.20), $\dot{V}$ can be upper bounded as follows

$$\begin{aligned} \dot{V} = \dot{V}_o + \varrho \dot{V}_c &\leq -\zeta_o \|\widetilde{\boldsymbol{x}}\|^2 - \varrho \zeta_c \|\boldsymbol{\varepsilon}\|^2 + \varrho \zeta_{co} \|\widetilde{\boldsymbol{x}}\| \|\boldsymbol{\varepsilon}\| \\ &= - \begin{bmatrix} \|\widetilde{\boldsymbol{x}}\| \\ \|\boldsymbol{\varepsilon}\| \end{bmatrix}^{\mathrm{T}} \begin{bmatrix} \zeta_o & -\varrho \zeta_{co}/2 \\ -\varrho \zeta_{co}/2 & \varrho \zeta_c \end{bmatrix} \begin{bmatrix} \|\widetilde{\boldsymbol{x}}\| \\ \|\boldsymbol{\varepsilon}\| \end{bmatrix}. \end{aligned} \tag{A.21}$$

The function $\dot{V}$ is guaranteed to be negative semi-definite if the arbitrary positive constant ϱ is chosen so as to satisfy the inequality

$$\varrho < \frac{4 \zeta_o \zeta_c}{\zeta_{co}^2}. \tag{A.22}$$

This guarantees the boundedness of all error signals. By invoking Barbalat's lemma [3], it can be recognized that $\dot{V} \to 0$, which implies the convergence to $\boldsymbol{0}$ of both $\widetilde{\boldsymbol{x}}$ and $\boldsymbol{\varepsilon}$, while the estimation errors $\widetilde{\theta}_c$ and $\widetilde{\theta}_o$ are only guaranteed to be uniformly bounded. □

A.4 Proof of Theorem 5.4

The closed-loop dynamics can be derived by plugging (5.45) and (5.46) into (5.15) and taking into account that $y_2 = y_{2,\mathrm{des}} - e_2$ and $\widetilde{\theta}_c = \theta - \widehat{\theta}_c$ (and thus, $\theta = \widehat{\theta}_c + \widetilde{\theta}_c$):

$$\dot{\boldsymbol{e}} = \boldsymbol{A}_c' \boldsymbol{e} + \boldsymbol{\psi}_c(\boldsymbol{y}) \widetilde{\theta}_c - \boldsymbol{A}_{co}'(\boldsymbol{y}) \widetilde{\boldsymbol{x}}, \tag{A.23}$$

where

$$\boldsymbol{A}_c' = \begin{bmatrix} -g_{\mathrm{P,r}} & \alpha_r \theta \\ 0 & -g_{\mathrm{P,j}} \end{bmatrix}, \qquad \boldsymbol{A}_{co}' = \begin{bmatrix} \boldsymbol{A}_{\mathrm{M,E}} & \boldsymbol{O}_{2\times 2} \end{bmatrix}.$$

Let us consider the following positive definite Lyapunov candidate scalar function:

$$V(\widetilde{\boldsymbol{x}}, \boldsymbol{e}, \widetilde{\theta}_o, \widetilde{\theta}_c) = V_o(\widetilde{\boldsymbol{x}}, \widetilde{\theta}_o) + \varrho V_c(\boldsymbol{e}, \widetilde{\theta}_c),$$

where V_o is the same function defined in (A.2), V_c is given by

$$V_c(\boldsymbol{e}, \widetilde{\theta}_c) = \frac{1}{2} \boldsymbol{e}^{\mathrm{T}} \boldsymbol{e} + \frac{1}{2} \gamma_c \widetilde{\theta}_c^2, \tag{A.24}$$

and $\varrho > 0$ is a positive constant to be determined.

The derivative of V_c along the trajectories of the system (A.1), (A.23), taking into account that the parameter θ is constant, is given by

$$\dot{V}_c = \boldsymbol{e}^T \boldsymbol{A}'_c \boldsymbol{e} - \boldsymbol{e}^T \boldsymbol{A}'_{co} \widetilde{\boldsymbol{x}} + \boldsymbol{\psi}_c^T \boldsymbol{e} \widetilde{\theta}_c - \gamma_c \dot{\widehat{\theta}}_c \widetilde{\theta}_c.$$

Taking into account the update law (5.47), $\dot{V}_c$ becomes

$$\dot{V}_c = \boldsymbol{e}^T \boldsymbol{A}'_c \boldsymbol{e} - \boldsymbol{e}^T \boldsymbol{A}'_{co} \widetilde{\boldsymbol{x}}.$$

Hence, $\dot{V}_c$ can be upper bounded as follows:

$$\dot{V}_c \leq -\zeta_c \|\boldsymbol{e}\|^2 + \zeta_{c,o} \|\widetilde{\boldsymbol{x}}\| \|\boldsymbol{e}\|, \tag{A.25}$$

where

$$\zeta_c = \min\{g_{P,r}, g_{P,j}\} \quad \text{and} \quad \zeta_{c,o} = \max_{i=1,\ldots,N_C} \{\overline{a}_i\}$$

with $\overline{a}_i$ defined in Appendix A.1, (A.4).

Hence, by considering the inequalities in (A.9) and (A.25), $\dot{V}$ can be upper bounded as follows:

$$\begin{aligned}\dot{V} = \dot{V}_o + \varrho \dot{V}_c &\leq -\zeta_o \|\widetilde{\boldsymbol{x}}\|^2 - \varrho \zeta_c \|\boldsymbol{e}\|^2 + \varrho \zeta_{c,o} \|\widetilde{\boldsymbol{x}}\| \|\boldsymbol{e}\| \\ &= -\begin{bmatrix} \|\widetilde{\boldsymbol{x}}\| \\ \|\boldsymbol{e}\| \end{bmatrix}^T \begin{bmatrix} \zeta_o & -\varrho \zeta_{c,o}/2 \\ -\varrho \zeta_{c,o}/2 & \varrho \zeta_c \end{bmatrix} \begin{bmatrix} \|\widetilde{\boldsymbol{x}}\| \\ \|\boldsymbol{e}\| \end{bmatrix}.\end{aligned}$$

The function $\dot{V}$ is guaranteed to be negative semi-definite if the arbitrary positive constant ϱ is chosen so as to satisfy the inequality

$$\varrho < \frac{4\zeta_o \zeta_c}{\zeta_{c,o}^2}.$$

This guarantees the boundedness of all error signals. By invoking Barbalat's lemma [3], it can be recognized that $\dot{V} \to 0$, which implies the global convergence to $\boldsymbol{0}$ of both $\widetilde{\boldsymbol{x}}$ and $\boldsymbol{e}$, while the parameters estimation errors $\widetilde{\theta}_o$ and $\widetilde{\theta}_c$ are only guaranteed to be uniformly bounded. □

A.5 Proof of Theorem 6.1

Let us consider the following positive definite Lyapunov candidate function:

$$V_{SMi}(\widetilde{\boldsymbol{x}}_{SMi}) = \frac{1}{2} \widetilde{\boldsymbol{x}}_{SMi}^T \boldsymbol{P}_o \widetilde{\boldsymbol{x}}_{SMi}, \tag{A.26}$$

where $\boldsymbol{P}_o$ is the same matrix defined in (A.3).

The derivative of $V_{\mathrm{SM}i}$ along the trajectories of the error dynamics (6.11), in the absence of uncertainties and noise (i.e., $\boldsymbol{\eta} = \mathbf{0}$ and $\boldsymbol{n} = \mathbf{0}$), is given by

$$
\begin{aligned}
\dot{V}_{\mathrm{SM}i} = & -\sum_{l=1}^{N_\mathrm{C}} \sigma_l k_{\mathrm{c}l} \widetilde{x}_{\mathrm{SM}i,l}^2 - (l_\mathrm{r} + \alpha_\mathrm{r}\theta)\widetilde{x}_{\mathrm{SM}i,N_\mathrm{C}+1}^2 - (l_\mathrm{j} + \alpha_\mathrm{j}\theta)\widetilde{x}_{\mathrm{SM}i,N_\mathrm{C}+2}^2 \\
& + \sum_{l=1}^{N_\mathrm{C}-1} \sum_{h=l+1}^{N_\mathrm{C}} \sigma_h \upsilon_{1,h} k_{\mathrm{c}l,h} \widetilde{x}_{\mathrm{SM}i,l} \widetilde{x}_{\mathrm{SM}i,h} + \sum_{l=1}^{N_\mathrm{C}} (a_l - \sigma_l l_l) \widetilde{x}_{\mathrm{SM}i,l} \widetilde{x}_{\mathrm{SM}i,N_\mathrm{C}+1} \\
& + (\alpha_\mathrm{r}\theta + \alpha_\mathrm{j}\theta)\widetilde{x}_{\mathrm{SM}i,N_\mathrm{C}+1}\widetilde{x}_{\mathrm{SM}i,N_\mathrm{C}+2},
\end{aligned}
\tag{A.27}
$$

where the dependence of the rate constants upon the temperature has been dropped for notation compactness. By considering the inequalities in (2.32) and (2.33) $\dot{V}_{\mathrm{SM}i}$ can be bounded as follows:

$$
\begin{aligned}
\dot{V}_{\mathrm{SM}i} \le & -\sum_{l=1}^{N_\mathrm{C}} \sigma_l \underline{k}_{\mathrm{c}l} \widetilde{x}_{\mathrm{SM}i,l}^2 - (l_\mathrm{r} + \alpha_\mathrm{r}\theta)\widetilde{x}_{\mathrm{SM}i,N_\mathrm{C}+1}^2 - (l_\mathrm{j} + \alpha_\mathrm{j}\theta)\widetilde{x}_{\mathrm{SM}i,N_\mathrm{C}+2}^2 \\
& + \sum_{l=1}^{N_\mathrm{C}-1} \sum_{h=l+1}^{N_\mathrm{C}} \sigma_h \upsilon_{l,h} \overline{k}_{\mathrm{c}i,h} |\widetilde{x}_{\mathrm{SM}i,l}| |\widetilde{x}_{\mathrm{SM}i,h}| \\
& + \sum_{l=1}^{N_\mathrm{C}} (\overline{a}_l + \sigma_l l_l) |\widetilde{x}_{\mathrm{SM}i,l}| |\widetilde{x}_{\mathrm{SM}i,N_\mathrm{C}+1}| \\
& + (\alpha_\mathrm{r}\theta + \alpha_\mathrm{j}\theta) |\widetilde{x}_{\mathrm{SM}i,N_\mathrm{C}+1}| |\widetilde{x}_{\mathrm{SM}i,N_\mathrm{C}+2}| \\
= & -\sum_{l=1}^{N_\mathrm{C}-1} \sum_{h=l+1}^{N_\mathrm{C}} \begin{bmatrix} |\widetilde{x}_{\mathrm{SM}i,l}| \\ |\widetilde{x}_{\mathrm{SM}i,h}| \end{bmatrix}^\mathrm{T} \boldsymbol{\Omega}_{l,h} \begin{bmatrix} |\widetilde{x}_{\mathrm{SM}i,l}| \\ |\widetilde{x}_{\mathrm{SM}i,h}| \end{bmatrix} \\
& - \sum_{l=1}^{N_\mathrm{C}} \begin{bmatrix} |\widetilde{x}_{\mathrm{SM}i,l}| \\ |\widetilde{x}_{\mathrm{SM}i,N_\mathrm{C}+1}| \end{bmatrix}^\mathrm{T} \boldsymbol{\Phi}_l \begin{bmatrix} |\widetilde{x}_{\mathrm{SM}i,l}| \\ |\widetilde{x}_{\mathrm{SM}i,N_\mathrm{C}+1}| \end{bmatrix} \\
& - \begin{bmatrix} |\widetilde{x}_{\mathrm{SM}i,N_\mathrm{C}+1}| \\ |\widetilde{x}_{\mathrm{SM}i,N_\mathrm{C}+2}| \end{bmatrix} \boldsymbol{\Pi} \begin{bmatrix} |\widetilde{x}_{\mathrm{SM}i,N_\mathrm{C}+1}| \\ |\widetilde{x}_{\mathrm{SM}i,N_\mathrm{C}+2}| \end{bmatrix},
\end{aligned}
$$

where $\overline{a}_i$ and the matrices $\boldsymbol{\Omega}_{l,h}$ and $\boldsymbol{\Phi}_l$ have been defined in Appendix A.1 ((A.4) and (A.5), respectively), while

$$
\boldsymbol{\Pi} = \begin{bmatrix} \alpha_\mathrm{r}\theta & -\frac{(\alpha_\mathrm{r}+\alpha_\mathrm{j})\theta}{2} \\ -\frac{(\alpha_\mathrm{r}+\alpha_\mathrm{j})\theta}{2} & l_\mathrm{j} + \alpha_\mathrm{j}\theta \end{bmatrix}.
$$

As shown in Appendix A.1, the matrices $\boldsymbol{\Omega}_{l,h}$ and $\boldsymbol{\Phi}_l$ are positive definite, provided that the gain l_r and the constants σ_ls satisfy inequalities (A.6) and (A.7), respectively. The matrix $\boldsymbol{\Pi}$ is positive definite if l_{j} satisfies the following inequality:

$$l_{\mathrm{j}} > \frac{(\alpha_{\mathrm{r}}^2 + \alpha_{\mathrm{j}}^2 + \alpha_{\mathrm{r}}\alpha_{\mathrm{j}})\theta}{4\alpha_r}. \tag{A.28}$$

Therefore, $\dot{V}_{\mathrm{SM}i}$ can be upper bounded as follows:

$$\begin{aligned}\dot{V}_{\mathrm{SM}i} \le &-\sum_{l=1}^{N_{\mathrm{C}}-1}\sum_{h=l+1}^{N_{\mathrm{C}}} \lambda_{\min}(\boldsymbol{\Omega}_{l,h})\big(\widetilde{x}_{\mathrm{SM}i,l}^2 + \widetilde{x}_{\mathrm{SM}i,h}^2\big)\\ &-\sum_{l=1}^{N_{\mathrm{C}}} \lambda_{\min}(\boldsymbol{\Phi}_l)\big(\widetilde{x}_{\mathrm{SM}i,l}^2 + \widetilde{x}_{\mathrm{SM}i,N_{\mathrm{C}}+1}^2\big)\\ &-\lambda_{\min}(\boldsymbol{\Pi})\big(\widetilde{x}_{\mathrm{SM}i,N_{\mathrm{C}}+1}^2 + \widetilde{x}_{\mathrm{SM}i,N_{\mathrm{C}}+2}^2\big).\end{aligned}$$

Thus,

$$\dot{V}_{\mathrm{SM}i} \le -\zeta_{\mathrm{SM}i}\|\widetilde{\boldsymbol{x}}_{\mathrm{SM}i}\|^2, \tag{A.29}$$

where

$$\begin{aligned}\zeta_{\mathrm{SM}i} &= \min\big\{(N_{\mathrm{C}}-1)\underline{\omega} + \underline{\phi},\, N_{\mathrm{C}}\underline{\phi},\, \lambda_{\min}(\boldsymbol{\Pi})\big\},\\ \underline{\omega} &= \min_{\substack{l=1,\dots,N_{\mathrm{C}}-1\\ h=l+1,\dots,N_{\mathrm{C}}}}\big\{\lambda_{\min}(\boldsymbol{\Omega}_{l,h})\big\},\quad \underline{\phi} = \min_{l=1,\dots,N_{\mathrm{C}}}\big\{\lambda_{\min}(\boldsymbol{\Phi}_l)\big\}.\end{aligned} \tag{A.30}$$

Hence, $\dot{V}_{\mathrm{SM}i}$ is negative semi-definite: this guarantees the boundedness of $\widetilde{\boldsymbol{x}}_{\mathrm{SM}i}$. By invoking Barbalat's lemma [3], it can be recognized that $\dot{V}_{\mathrm{SM}i} \to 0$, which implies the global uniform convergence to $\mathbf{0}$ of $\widetilde{\boldsymbol{x}}_{\mathrm{SM}i}$ as $t \to \infty$. □

A.6 Proof of Theorem 6.2

Let us consider the following positive definite Lyapunov candidate function:

$$V_i\big(\widetilde{\boldsymbol{x}}_i, \widetilde{\boldsymbol{\theta}}_{\mathrm{f},i}\big) = \frac{1}{2}\widetilde{\boldsymbol{x}}_i^{\mathrm{T}}\boldsymbol{P}_{\mathrm{o}}\widetilde{\boldsymbol{x}}_i + \frac{1}{2}\gamma_i\widetilde{\boldsymbol{\theta}}_{\mathrm{f},i}^{\mathrm{T}}\widetilde{\boldsymbol{\theta}}_{\mathrm{f},i}, \tag{A.31}$$

where $\boldsymbol{P}_{\mathrm{o}}$ is the same matrix defined in (A.3). The derivative of V_i along the trajectories of the error dynamics (6.29), in the absence of uncertainties and noise (i.e., $\boldsymbol{\eta} = \mathbf{0}$ and $\boldsymbol{n} = \mathbf{0}$) and taking into account that the parameter vector $\boldsymbol{\theta}_{\mathrm{f}}$ is constant, is given by

$$
\begin{aligned}
\dot{V}_i = & -\sum_{l=1}^{N_C}\sigma_l k_{cl}\widetilde{x}_{i,l}^2 - (l_r + \alpha_r\theta)\widetilde{x}_{i,N_C+1}^2 - (l_j + \alpha_j\theta)\widetilde{x}_{i,N_C+2}^2 \\
& + \sum_{l=1}^{N_C-1}\sum_{h=l+1}^{N_C}\sigma_h \upsilon_{l,h} k_{cl,h}\widetilde{x}_{i,l}\widetilde{x}_{i,h} + \sum_{l=1}^{N_C}(a_l - \sigma_l l_l)\widetilde{x}_{i,l}\widetilde{x}_{i,N_C+1} \\
& + (\alpha_r\theta + \alpha_j\theta)\widetilde{x}_{i,N_C+1}\widetilde{x}_{i,N_C+2} + \widetilde{\boldsymbol{x}}_i^{\mathrm{T}}\boldsymbol{C}^{\mathrm{T}}\boldsymbol{\varphi}_i(t,\boldsymbol{y},u)\widetilde{\boldsymbol{\theta}}_{\mathrm{f},i} - \gamma_i\dot{\widehat{\boldsymbol{\theta}}}{}_{\mathrm{f},i}^{\mathrm{T}}\widetilde{\boldsymbol{\theta}}_{\mathrm{f},i},
\end{aligned}
$$

where the dependence of the rate constants upon the temperature has been dropped for notation compactness. By considering the update law (6.28), $\dot{V}_i$ has the same form of $\dot{V}_{\mathrm{SM}i}$ (see (A.27)), thus the proof follows the same steps of the proof of Theorem 6.1, and it is possible to write

$$
\dot{V}_i \le -\zeta_i\|\widetilde{\boldsymbol{x}}_i\|^2, \tag{A.32}
$$

where ζ_i has the same expression of $\zeta_{\mathrm{SM}i}$ in (A.30).

Hence, $\dot{V}_i$ is negative semi-definite: this guarantees the boundedness of $\widetilde{\boldsymbol{x}}_i$ and $\widetilde{\boldsymbol{\theta}}_{\mathrm{f},i}$. By invoking Barbalat's lemma [3], it can be recognized that $\dot{V}_i \to 0$, which implies the global uniform convergence to $\mathbf{0}$ of $\widetilde{\boldsymbol{x}}_i$ as $t \to \infty$, while $\widetilde{\boldsymbol{\theta}}_{\mathrm{f},i}$ is only guaranteed to be uniformly bounded.

The above result is usual in direct adaptive estimation and/or control schemes. The exponential convergence to zero of both the state estimation error and the parameter estimation error is guaranteed only in the presence of the persistency of excitation condition [1, 3]. This, in turn, implies that they keep bounded in the presence of bounded uncertainties. However, since persistency of excitation may be difficult to guarantee in practice, a modified parameters update law can be adopted, according to the concept of projection operator [2]. In detail, adoption of the following update law instead of (6.28),

$$
\dot{\widehat{\boldsymbol{\theta}}}_{\mathrm{f},i} = \left(\frac{\boldsymbol{I}_{m_i} - \Lambda\gamma_i^{-1}\widehat{\boldsymbol{\theta}}_{\mathrm{f},i}\widehat{\boldsymbol{\theta}}_{\mathrm{f},i}^{\mathrm{T}}}{\|\widehat{\boldsymbol{\theta}}_{\mathrm{f},i}\|}\right)\gamma_i^{-1}\boldsymbol{\varphi}_i^{\mathrm{T}}(\boldsymbol{y},u)\widetilde{\boldsymbol{y}}_i, \tag{A.33}
$$

where Λ is a boolean variable defined as

$$
\Lambda = \begin{cases} 0, & \text{if } \{\|\widehat{\boldsymbol{\theta}}_{\mathrm{f},i}\| < M\} \text{ or } \{\|\widehat{\boldsymbol{\theta}}_{\mathrm{f},i}\| = M \text{ and } \gamma_i^{-1}\boldsymbol{\varphi}_i^{\mathrm{T}}(\boldsymbol{y},u)\widetilde{\boldsymbol{y}}_i \le 0\}, \\ 1, & \text{if } \{\|\widehat{\boldsymbol{\theta}}_{\mathrm{f},i}\| = M \text{ and } \gamma_i^{-1}\boldsymbol{\varphi}_i^{\mathrm{T}}(\boldsymbol{y},u)\widetilde{\boldsymbol{y}}_i > 0\}, \end{cases}
$$

guarantees that $\widehat{\boldsymbol{\theta}}_{\mathrm{f},i}$ never leaves the hypersphere

$$
\mathcal{M} = \left\{\widehat{\boldsymbol{\theta}}_{\mathrm{f},i} \in \mathbb{R}^{m_i} : \left\|\widehat{\boldsymbol{\theta}}_{\mathrm{f},i}\right\| \le M\right\},
$$

and thus $\widetilde{\boldsymbol{\theta}}_{\mathrm{f},i}$ keeps uniformly bounded in the presence of uncertainties as well. □

References

1. K.J. Aström and B. Wittenmark. *Adaptive Control, 2nd Edition*. Addison-Wesley, Reading, 1995.
2. P.A. Ioannou and J. Sun. *Robust Adaptive Control*. Prentice Hall, Upper Saddle River, 1996.
3. H.K. Khalil. *Nonlinear Systems, 2nd Edition*. Prentice Hall, Upper Saddle River, 1996.

Index

F. Caccavale et al., *Control and Monitoring of Chemical Batch Reactors*,
Advances in Industrial Control,
DOI 10.1007/978-0-85729-195-0,